structural masonry

Sven Sahlin

Professor of Structural Mechanics
Chalmers University of Technology
Göteborg, Sweden

Prentice-Hall, Inc.
Englewood Cliffs, New Jersey

PRENTICE-HALL INTERNATIONAL, INC., *London*
PRENTICE-HALL OF AUSTRALIA, PTY. LTD., *Sydney*
PRENTICE-HALL OF CANADA, LTD., *Toronto*
PRENTICE-HALL OF INDIA PRIVATE LIMITED., *New Delhi*
PRENTICE-HALL OF JAPAN, INC., *Tokyo*

to kerstin

foreword

This monograph is the first comprehensive treatise in English on the behavior of structural masonry. It is more than a collection of facts and data: it gives the reader an understanding of the physical behavior of structural masonry, and a rational theoretical basis for predicting the behavior of elements or structures composed of masonry units.

Dr. Sahlin has been working in this field for many years in Sweden, and in 1967–68 he was on our staff at the University of Illinois, where the notes for this book were prepared. Preliminary reviews of his manuscript by other experts in the field indicate that the author has done a masterful job in collecting and coordinating the information available and in synthesizing it for ready use by students, research workers, and structural engineers and designers.

We are pleased that Dr. Sahlin has devoted his time and energy to the preparation of this monograph, and we feel sure it will prove to be most useful and valuable in the field of structural masonry design.

N. M. NEWMARK AND W. J. HALL

introduction

This book has been prepared for students and for practicing engineers who wish to acquire knowledge of the behavior of structural masonry. It can be used as a textbook; in fact, the first draft for the manuscript was a set of notes prepared for a graduate course at the University of Illinois, in Urbana. The book also contains material of value to the practicing engineer, such as design methods, strength properties, and case studies. Some of the material was originally collected when the writer participated in the preparation of that portion of the Swedish Building Codes which deals with masonry structures, so the book should also be of interest building authorities.

The writer has applied general theories of structural mechanics to the analysis of masonry as far as possible, and has illustrated the possibility of using common structuring design methods for masonry. In situations where masonry has a behavior different from elastic structural materials, the differences have been pointed out and the actual phenomena explained in detail. Although there are differences, under certain circumstances, between different types of masonry, the similarities have been emphasized, since the same phenomena often govern behavior under load, regardless of the material in the units.

The writer has felt this approach important, since most of the readily available literature written by manufacturing associations, etc., naturally treats only one material, and the user is required to collate unconnected literature to cover the field. Besides, hardly any monographs exist today covering the common masonry types.

The subject of the book is the structural use of masonry. The main portion of the book is therefore devoted to design and estimate of the strength of masonry under common basic loading conditions and under combined loadings. However, a few chapters dealing with material properties are also included, as well as chapters on cracking, earth pressure against foundation walls, practical applications, and case studies.

SVEN SAHLIN

acknowledgments

Most of the manuscript for this book was written at the Civil Engineering Department of the University of Illinois, in Urbana. I am grateful to Dr. Nathan N. Newmark, Chairman of the Department, for giving me the opportunity to visit the Department for one year, thereby giving me time to work on the manuscript for this book. The stay at the Department was combined with a sabbatical leave from the Royal Institute of Technology in Stockholm.

Some of the research and test results included in the book are taken from earlier work carried out on different occasions at the Division of Building Statics and Structural Engineering in the Royal Institute of Technology in Stockholm, in cooperation with the Swedish brick and block manufacturers.

I am particularly indebted to Dr. M. A. Sozen, Professor of Civil Engineering at the University of Illinois, who was instrumental in the underaking, who read the manuscript and suggested improvements, who let me benefit from participation in some of his research (particularly on filler in walls), and who constantly encouraged me in my work.

Several other persons have given advice and assistance in the preparation of this book. The counsel of Dr. Hubert K. Hilsdorf, Professor of Civil Engineering at the University of Illinois, who also read the manuscript and suggested improvements, is appreciated. Civilingenjör Hans Falk and Tekniske Licens Bo-Göran Hellers at the Royal Institute of Technology have been of assistance in furnishing information and literature. Mr. Hellers also read the manuscript and suggested improvements. The Tegelindustrins Centralkontor AB (Swedish Brick Manufacturers' Association) and AB Lättbetong (Light Weight Cellular Concrete Company) have willingly provided information and literature. SCPI, PCA, and NCMA have assisted in several ways.

The typing of the manuscript—which represents, essentially, the notes used in a course given at the University of Illinois in the spring of 1968—has been done by the staff at the University of Illinois: Mrs. Bonnie Webster, Mrs. Ruth Worner, Mrs. Deanle McMaster, and Mrs. Nancy Segall.

The invaluable help received from the editors, Dr. Nathan M. Newmark and Dr. William J. Hall, as well as from the publisher, is appreciated. Dr. Hall also reviewed the manuscript and suggested alterations and improvements.

contents

A ☐ Masonry units: bricks, tiles, and blocks

A.1 ☐ Introductory remarks

Masonry is normally laid of rectangular units of different materials, shapes, and sizes. The common types of units are clay bricks, clay tiles, concrete blocks, light weight cellular concrete blocks, sand–lime bricks, and natural building stones. The different types of masonry units are briefly described in what follows.

Building *clay bricks* are normally rectangular masonry units made of clay, shale, fireclay, or mixtures thereof by heating (baking or burning) at 750 to 1300°C (1400 to 2370°F). During the heating period (when the temperature rises), water is driven off and combustible material is oxidized or burned off. Finally, the stage is reached when vitrification takes place and the grains fuse together (to different degrees, depending on the temperature).

The raw material used has the approximate composition illustrated in Fig. A.1 [A.1].† A more elaborate definition and classification of bricks (for legal and technical use) can be found in ASTM, Part 12 [A.2], especially in Sections C34, C62, C212, and C216.

Bricks are (or have been) available in a large variation of shapes and sizes, from approximately $12 \times 9 \times 4.5$ cm ($4.5 \times 3.4 \times 1.7$ in.) to $60 \times 60 \times 60$ cm ($2 \times 2 \times 2$ ft) (Plummer and Reardon [A.3]), with the modular sizes shown in Table A.1 [A.4] currently common in the United States. The actual sizes of the bricks are the nominal sizes minus the joint thickness. Examples are shown in Fig. A.2.

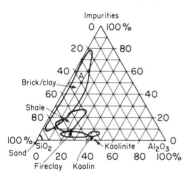

Figure A.1 ☐ Composition of raw materials for clay products [A.1].

†Numbers in brackets refer to the references at the end of each chapter.

1

Examples of solid structural clay bricks: Examples of structural clay tiles:

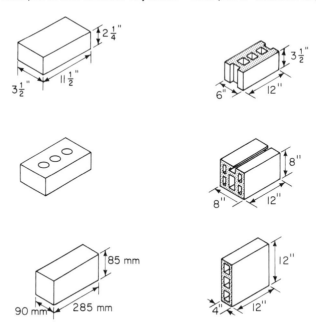

Figure A.2a ☐ *Sizes and shapes of structural clay units [A.4].*

Clay bricks can be solid or hollow. In the United States they are defined as solid if the net cross-sectional area in every plane parallel to the bearing surface is 75% or more of its gross cross-sectional area measured in the same plane; they are defined as hollow if the cores, cells, or hollow spaces within the total cross-sectional area exceed 25% of the cross section of the unit. The cores can have different shapes, and over 100 cores in a brick have been used. A few examples of bricks are shown in Fig. A.2. The density of solid clay bricks ranges from 1300 to 2200 kg/m³ (80 to 140 lb/ft³).

Clay tiles are hollow burned clay, shale, fireclay, or mixtures thereof; units with the typical shapes shown in Fig A.2. They

Figure A.2b ☐ *Typical shapes and sizes of concrete masonry units. Dimensions shown are actual unit sizes: a $7\frac{5}{8} \times 7\frac{5}{8} \times 15\frac{5}{8}$-inch unit is commonly known as an $8 \times 8 \times 16$-inch concrete block. Half-length units are usually available for most of the units shown. See concrete products manufacturers for shapes and sizes of units available locally [A.5].*

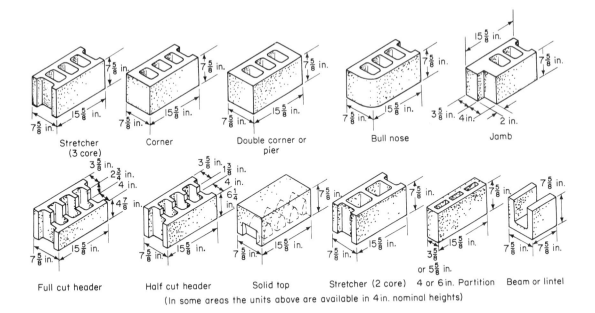

Stretcher (3 core) Corner Double corner or pier Bull nose Jamb

Full cut header Half cut header Solid top Stretcher (2 core) 4 or 6 in. Partition Beam or lintel

(In some areas the units above are available in 4 in. nominal heights)

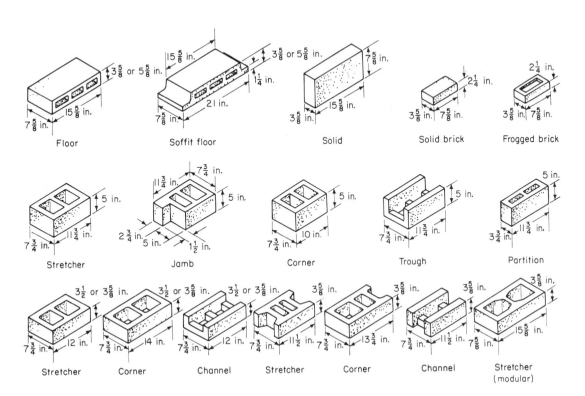

Floor Soffit floor Solid Solid brick Frogged brick

Stretcher Jamb Corner Trough Partition

Stretcher Corner Channel Stretcher Corner Channel Stretcher (modular)

Notations in Chapter A

f' = compressive strength
IRA = initial rate of absorption (defined in ASTM 67) g/min per 30 sq in.
S = suction (g/min per dm^2)

have relatively thin webs and shells. The cells can be parallel to the main loading direction, giving a wall with vertical cells, or perpendicular to the main loading direction, giving a wall with horizontal cells. The main reason for making the units hollow is to decrease material consumption and weight. Hollow units also have better heat insulation.

Concrete masonry building units or blocks are made of normal or light weight aggregate concrete in many types and sizes, some of which are shown in Fig. A.2 [A.5]. The units are molded of a mixture of sand, gravel (or crushed stone; aircooled, iron blast furnace slag; pumice; scoria; tuff; cinders; or expanded clay, shale, slate, diatomite, fly ash, perlite, and vermiculite), cement, and water in a machine three or more units at a time under pressure and/or vibration. The blocks are often steamcured and autoclaved (ASTM, part 12 [A.2] and ACI committee 516 [A.19]),

If the units have more than 25% coring, they are classified as hollow. The density of concrete blocks with 40% core area is about 90 lb/ft^3 (1500 kg/m^3)(Kuenning and Carlson [A.6]). With light weight aggregates (ASTM, C331 [A.1]) such as expanded

Table A.1 ☐ Nominal modular sizes of some bricks [A.4].

Unit designation	Thickness (inches)	Face dimension	
		Height (inches)	Length (inches)
Conventional brick	4	$2\frac{2}{3}$	8
Roman brick	4	2	12
Norman brick	4	$2\frac{2}{3}$	12
Engineer's brick	4	$3\frac{1}{5}$	8
Economy brick	4	4	8
Jumbo brick	4	4	12
Double brick	4	$5\frac{1}{3}$	8
Triple brick	4	$5\frac{1}{3}$	12
"SCR brick"	6	$2\frac{1}{3}$	12

clay or shale, expanded slag or pumice, densities ranging from 20 to 120 1b/ft³ (400 to 1900 kg/m³) can be obtained, with usual values from 50 to 100 1b/ft³ (800 to 1600 kg/m³) [A.6], [A.7.].

Light weight cellular concrete blocks are generally produced by a hydraulic cementitious binder plus ground sand and are made porous by addition of chemicals (Al-powder). Casting and swelling is followed by cutting and usually by a high-pressure steam-curing process to harden and improve the strength and dimensional stability of the blocks. Volumetric weights of 400 to 700 kg/m³ (64 to 110 lb/ft³) are obtained by variations of the ingredients [A.8.]

Sand–lime building bricks are made of siliceous sand and hydrated high calcium lime under pressure and elevated temperature. The units are then autoclaved. Their volumetric weight is approximately 1800 to 2000 kg/m³ (110 to 125 lb/ft³).

Natural building stones can, of course, be found in any size and shape. Normally the weight must be limited for easy handling. When the stones are connected to other masonry units, precutting to standardized sizes is required, as for other units.

A.2 □ Strength

The strength properties that commonly are tested on *bricks* are the compressive strength and the modulus of rupture. The compressive strength test is made on a piece of brick of the same approximate length and width. The flatwise faces of the brick

Figure A.3a □ Compressive strength of bricks (flatwise). Histogram for 37 % of the total production of bricks in the United States in 1929 [A.9].

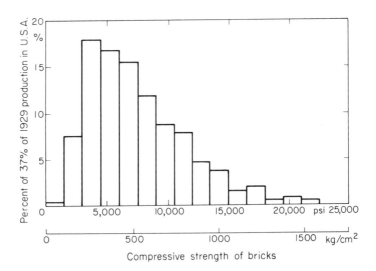

Compressive strength of bricks

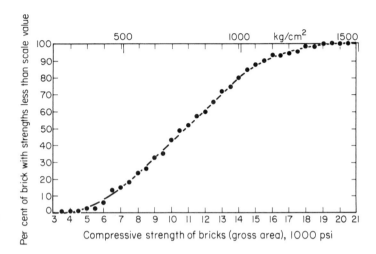

Figure A.3b □ Distribution function for compressive strengths of bricks produced in the United States [A.20].

are capped and, after aging, the bricks are compressed to failure in a testing machine. In some countries, two pieces of brick with a mortar joint are used as a test specimen. The modulus of rupture test is made by supporting a brick flatwise (on a span of 7 in. in the United States) and loading it at mid span to failure. Even this test procedure varies somewhat from country to country and depends on the brick size.

According to McBurney and Lovewell [A.9], the compressive strength of bricks produced in 1929 (37% of the total production in the United States) ranged up to approximately 22,000 psi (1550 kg/cm²); common values were 4000 to 10,000 psi (280 to 700 kg/cm²) (see Fig. A.3a). The values in Fig. A.3a correspond to a median strength of about 7000 psi (500 kg/cm²). A recent investigation (by Monk [A.20]) shows a median of over 10,500 psi (750 kg/cm²), as can be seen in Fig. A.3b. The modulus of rupture for bricks was up to 3450 psi (240 kg/cm²) according to the 1929 investigation, shown in Fig. A.4, with the normal values 400 to 2000 psi (25 to 150 kg/cm²) and the median about 1100 psi (75 kg/cm²).

If the change in modulus of rupture that has taken place over the years amounts to the same percentage as for the compressive strength, the median would be about 1600 psi (115 kg/cm²) in 1965.

Figures A.3a, A.3b, and A.4 give a good idea of diversity of brick strengths. The supply of the different strengths varies geographically since the locally available clay to a great extent determines the strength of the product.

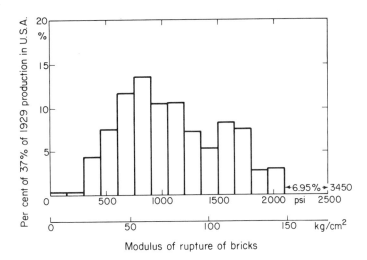

Figure A.4 □ Modulus of rupture of bricks. Histogram for 37 % of the total production of bricks in the United States in 1929 [A.9].

According to investigations described by Nevander [A.10], the modulus of rupture of solid brick varies within the limits 32 to 14% of their crushing strength for crushing strength values ranging from 194 to 506 kg/cm² (2750 to 7150 psi). This statement seems to be in reasonable agreement with Figs. A.3 and A.4, if strengths which occur in more than, say, 7% of the chosen fractions of production are considered.

For one specific brick quality, the splitting (tensile) strength was about 8% of the compressive strength for bricks with a compressive strength of 364 kg/cm² (5200 psi), according to Hilsdorf [A.11].

The ratio of modulus of rupture to compressive strength was 0.11, 0.15, and 0.17 for three uncored solid tested bricks (Hilsdorf [A.11]). For three tested types of cored bricks, the averages were 0.11, 0.10, and 0.10 (SCPRF [A.12]). The standard deviation of the observed strengths was about 25%, according to Hilsdorf [A. 11], and 5 to 20% according to SCPRF [A.12].

The compressive strength of *concrete* hollow *blocks* is about 35 to 200 kg/cm² (500 to 3000 psi) after 28 days, based on the gross area (three-hole units) [A.6], [A.7], [A.13]. With different types of light weight aggregates, values from 40 to 80 kg/cm² (550 to 1140 psi) were obtained, according to Richart, Moorman, and Woodworth [A.7], for blocks with densities of 750 to 1350 kg/m³ (47 to 84 lb/ft³).

The compressive strength of concrete masonry blocks is not only a function of the unit weight, aggregate type, and w/c (water to cement) ratios but also of the curing process; thus,

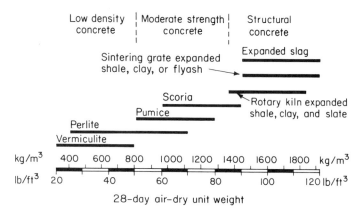

Figure A.5 □ Approximate density and use classification of light weight aggregate concretes [A.14].

28-day air-dry unit weight

the user has to rely on tests on finished blocks from the actual manufacturer.

A comprehensive description of properties of light weight aggregate concrete is to be found in Figs. A.5, A.6, and A.7 (ACI [A.14]). Even though the requirements and mixes for light weight aggregate concrete are somewhat different for structural cast-in-place use than for the manufacture of concrete blocks, Figs. A.5, A.6, and A.7 give a good idea of the range of the different values obtainable for such properties as volumetric weight, strength, and modulus of elasticity. Since the blocks usually are cored (see Fig. A.2), the gross volumetric weight of a concrete masonry wall is lower than indicated for the specific type of concrete used in the blocks. Other properties are also adjusted according to the area or the volume, whichever applies. The modulus of rupture of the concrete in the blocks is about 20% of the compressive strength [A.6].

The shrinkage of concrete blocks, which sometimes causes severe cracking in masonry, can be reduced by precarbonation if the blocks are cured in steam at atmospheric pressure. Autoclaving reduces the carbonation shrinkage considerably (Shiedeler [A.15], Fig. A.8). See also ACI 516 [A.19]. The shrinkage of concrete blocks is also affected by the type of aggregate used (Fig. A.9, according to ACI Committee 716 [A.16]; see also Fig. J.8).

Figure A.6 □ (Right) Cement content, water content, density, modulus of elasticity, creep, and shrinkage of light weight concrete of different compressive strengths [A.14].

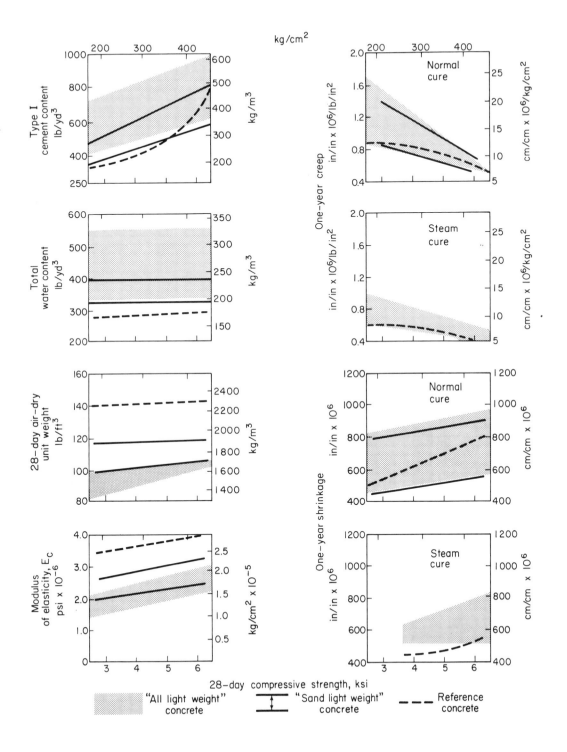

28-day compressive strength, ksi

"All light weight" concrete

"Sand light weight" concrete

Reference concrete

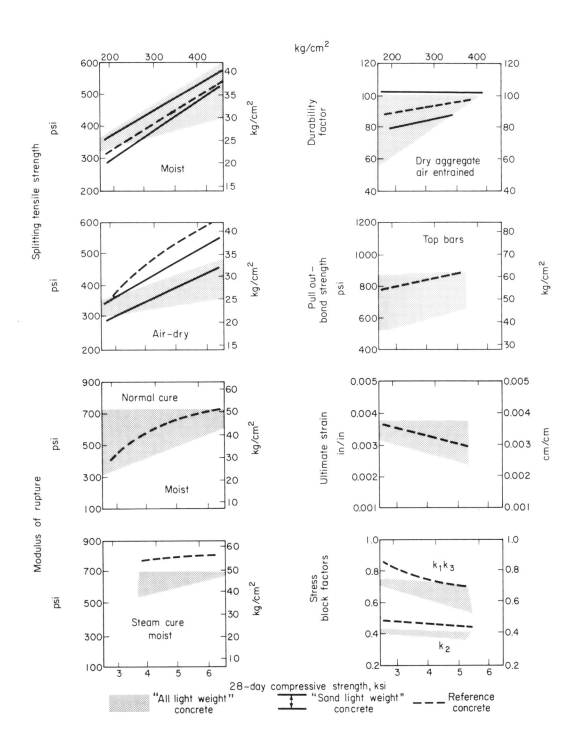

Figure A.7 ☐ *(Left) Splitting tensile strength, modulus of rupture, durability freezing and thawing, bond strength pullout tests, ultimate strain, and stress block factors of light weight concrete of different compressive strengths [A.14].*

Sand–lime building bricks have a compressive strength of approximately 75 to 1000 kg/cm² (1000 to 15,000 psi). The modulus of rupture is about 20% of the compressive strength (ASTM C73-51 [A.2], ACI 516 [A.19]).

Light weight cellular concrete blocks (density 300 to 800 kg/m³) have compressive strengths of 10 to 100 kg/cm² (150 to 1500 psi) and a modulus of rupture of approximately $\frac{1}{5}$ to $\frac{1}{3}$ of the compressive strength [A.8].

Natural building stones have compressive strengths from 100 to 4000 kg/cm² (1500 to 60,000 psi). The properties depend on volumetric weight, type of mineral, geological history, etc.

A.3 ☐ Modulus of elasticity

The modulus of elasticity of clay brick is roughly $300f'$ (Fig. D.4). The modulus of elasticity is approximately 500 to $1500f'$ for masonry concrete blocks (see Fig. A.10, based on data by Richart, Moorman, and Woodworth [A.7]; see also Fig. A.5, Eq. (D.23), and [A.6]).

The modulus of elasticity of light weight cellular concrete is about 15,000 kg/cm² (200,000 psi) for blocks with volumetric weight 500 kg/m³ (31 lb/ft³), 10,000 kg/cm² (140,000 psi), and 25,000 kg/cm² (350,000 psi) for volumetric weights 400 kg/m³ (25 lb/ft³) and 650 kg/m³ (40 lb/ft³), respectively [A.8].

Figure A.8 ☐ *(Below left) Shrinkage of block walls made with units with and without precarbonation, block cured at atmospheric pressure [A.15].*

Figure A.9 ☐ *(Below right) Relationship between shrinkage of concrete made with different aggregates and time (British method) [A.16].*

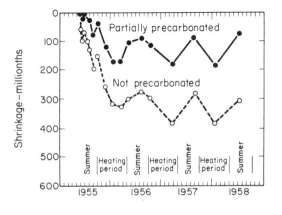

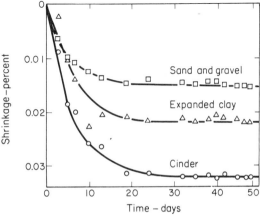

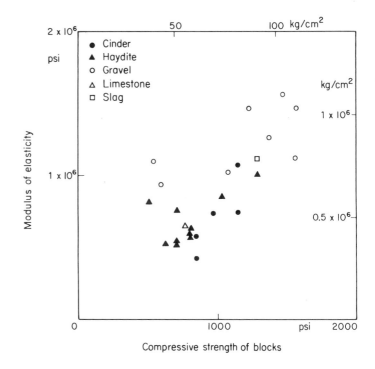

Figure A.10 ☐ Relationship between modulus of elasticity and compressive strength of masonry concrete blocks made of different aggregates [A.7]. Stresses and moduli are calculated on gross area.

The modulus of elasticity of natural building stones is approximately 4 to $9 \cdot 10^5$ kg/cm² (5.7 to $13.0 \cdot 10^6$ psi).

A.4 ☐ Initial rate of absorption (IRA) (suction)

Several different measurements of the absorption of bricks can be found in the literature: (1) initial rate of absorption (IRA), vaguely called *suction* (suction is in some countries measured somewhat differently, but the initial rate of absorption is defined in ASTM C67 [A.2]); (2) absorption in 24-hour submersion test; (3) absorption in 5-hour boiling test; (4) still other procedures not defind in ASTM. At worst the word *suction* is used without reference to any method of measurement.

Most of the different types of absorption tests give values with the same tendencies, since a masonry unit having a high water absorption during 1 minute also is likely to exhibit a high absorption under a prolonged period of immersion, as can be seen from Fig. A.11 (reported by Voss [A.17]).

Since IRA influences most of the masonry's properties, its definition will be briefly stated. According to ASTM 67 [A.2], the initial rate of absorption is measured as the amount of water initially absorbed by a dry unit when it is partially immersed in water to a depth of $\frac{1}{8}$ in. for a period of 1 minute. IRA is measured in grams per minute per 30 square inches. The value varies from 1 to 60 or even more. In European standards, the suggested test procedure is slightly different. The immersion depth is 1 cm (0.394 in.) and the suction S is measured in grams

Figure A.11 ☐ *Absorption curves for five different bricks [A.17].*

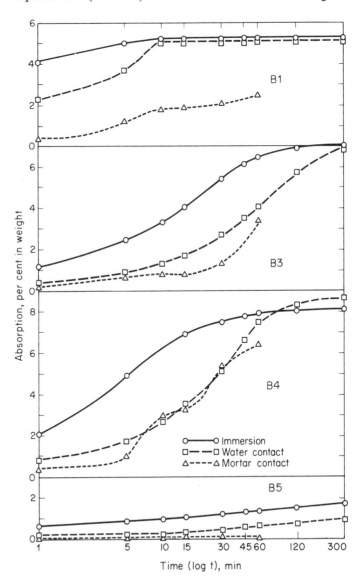

per minute per square decimeter (15.6 sq in.). The initial rate of absorption shows a strong relationship to strengths, especially on the modulus of rupture of masonry. The suction (IRA) varies inversely and the strength directly with the density of the brick, and the method for testing the effect of the suction (and for improving masonry in practice) is to decrease it by wetting the bricks and then measuring the improvement in strength of the masonry.

The 48-hour absorption for some bricks is plotted against the specific weight of the bricks in Fig. A.12. It is obvious that the denser masonry units have the least absorption.

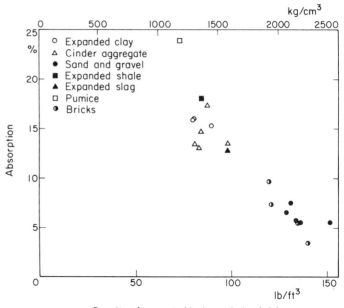

Figure A.12 ☐ Relationship between absorption (24 to 48 hr) and density of light weight concrete in concrete blocks and clay bricks [A.6], [A.12], [A.16].

Absorption by masonry concrete units is dependent on the specific weight of concrete, as shown in Fig. A.12 (data by ACI Committee 716 [A.16], and by Kuenning and Carlson [A.6]). In these cases the blocks were immersed in water for 24 hours. Concrete blocks should, however, never be wetted before laying because of the increase in shrinkage caused by the wetting.

The suction of sand–lime bricks is usually less than 20 g/dm² min (~ 35 g/30 sq in. min), but values between 5 and 50 g/dm² min have been observed [A.18].

References for Chapter A

A.1 ☐ Murphy, G.: "Properties of Engineering Materials," 2nd edition. International Textbook Company, Scranton, Pennsylvania, 1947.

A.2 ☐ American Society for Testing and Materials. Part 12: "Chemical-Resistant Nonmetallic Materials; Clay and Concrete Pipe and Tile; Masonry Mortars and Units; Asbestos–Cement Products; Natural Building Stones," ASTM Book of Standards, 1967.

A.3 ☐ Plummer, H., and Reardon, L.: "Principles of Brick Engineering (Handbook of Design)." Structural Clay Products Institute, Washington, D.C., 1939.

A.4 ☐ Caravaty, R., and Plummer, H.: "Principles of Clay Masonry Construction, Student's Manual," Structural Clay Products Institute, Washington, D.C., 1960.

A.5 ☐ National Concrete Masonry Association: "NCMA, TEK 2, AIA 10-C." Arlington, Virginia, 1965.

A.6 ☐ Kuenning, W. H., and Carlson, C. C.: "Effect of Variations in Curing and Drying on the Physical Properties of Concrete Masonry Units." Development Department Bulletin D13, Portland Cement Association, December, 1956.

A.7 ☐ Richart, F. E., Moorman, R. R., and Woodworth, P. M.: "Strength and Stability of Concrete Masonry Walls." Engineering Experiment Station Bulletin No. 251, University of Illinois, July 5, 1932.

A.8 ☐ Lättbetonghandboken (Handbook of Light Weight Cellular Concrete.) A. B. Lättbetong, Stockholm, 1965.

A.9 ☐ McBurney, J. W., and Lovewell C. E.: "Strength, Water Absorption and Weather Resistance of Building Bricks Produced in the United States." ASTM Proceedings of the Thirty-Sixth Annual Meeting, Vol. 33, Part II, Technical Papers, 1933.

A.10 ☐ Nevander, L. E.: "Provningar av Tegelmurverk," (Tests on Brick Walls.) Tegel, No. 5, Stockholm, 1954.

A.11 ☐ Hilsdorf, H. K.: "Untersuchungen über die Grundlagen der Mauerwerksfestigkeit." Bericht Nr. 40, Materialprüfungsamt für das Bauwesen der Technischen Hochschule, München, 1965.

A.12 ☐ Structural Clay Products Research Foundation. "Compressive, Transverse and Racking Strength Tests of Four-Inch Brick Walls." Research Report No. 9, Geneva, Ill., 1965.

A.13 ☐ Fishburn, C.: "Effect of Mortar Properties on Strength of Masonry." National Bureau of Standards, Monograph 36, Washington, D. C., 1961.

A.14 ☐ ACI Committee 213: "Guide for Structural Light Weight Aggregate Concrete." Journal of the American Concrete Institute, No. 8, Proceedings, Vol. 64, August, 1967.

A.15 ☐ Shiedeler, J. J.: "Carbonation Shrinkage of Concrete Masonry Units." PCA Bulletin D69, 1963.

A.16 ☐ ACI Committee 716 (W. C. Hansen, Chairman): "Physical Properties of High-Pressure Steam-Cured Block." Journal of the American Concrete Institute, Proceedings, Vol. 49, pp. 745–756, April, 1953.

A.17 ☐ Voss, W. C.: "Permeability of Brick Masonry Walls—An Hypothesis." ASTM Proceedings of the Thirty-Sixth Annual Meeting, Vol. 33, Part II, Technical Papers, 1933.

A.18 ☐ Dührkop, Saretok, Sneck, and Svendsen: "Bruk-Murning-Putsning." (Mortar-Masonry-Plastering.) National Swedish Council for Building Research, 1966.

A.19 ☐ ACI Committee 516. "High Pressure Steam Curing: Modern Practice and Properties of Autoclaved Products." Journal of the American Concrete Institute, No. 8, August, 1965.

A.20 ☐ Monk, Clarence B.: "A Historical Survey and Analysis of the Compressive Strength of Brick Masonry." Structural Clay Products Research Foundation. Research Report No. 12, Geneva, Ill., 1967.

B □ Masonry mortars

B.1 □ Introductory remarks

In order to ensure a good quality of masonry, a mortar must be equally well suited for both brick laying and load carrying when hardened. The workability of the fresh mortar must be such that the mason can fill all joints easily. When a course of units has been laid, the mortar and brick system must attain a reasonable rigidity before the next course is laid, in order to prevent excessive racking movements. On the other hand, if the mortar and brick system stiffens too fast, it can be impossible for the mason to make the necessary corrective movements of the newly laid unit.

The requirements mentioned above are not easily fulfilled with one single type of mortar since the properties of the units, especially the suction, also play an important role for the total result. It is possible that the ratio between the suction of the masonry units and the mortar's water retentivity may be such that only a thin mortar layer close to the units dries out—and, more specifically, dries out too fast. In such cases the bond between units and the mortar will suffer, and cracks will appear between the units and the hardened mortar joints. It it also possible that the water retentivity may be so small that the mortar is dried out by the lower unit, so that the new unit is laid in a partly dried and set mortar bed. In such cases the strength of the mortar joint will usually be low, and cracks will soon appear.

There is a relationship between the quality of the masonry and the suction of the units as well as the physical properties of the mortar. The numerical values of the factors which govern the relationship, however, are not well-known; in this area one must rely mainly on experience. Building standards and building codes embody empirical results which provide guiding data for proportioning and mixing of mortars (ASTM [B.1], [B.2], [B.3]). The properties of aggregates and binders (cementitious material) for mortars are discussed in the section which follows.

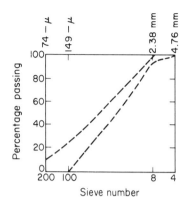

Figure B.1 □ *(Left) Limits for sieve curves for mortar sand, ASTM C144–66T [B.1]. In addition, the fineness modulus must be between 1.6 and 2.5 (percent retained on Nos. 100, 50, 30, and 16 divided by 100). Furthermore, the water demand ratio by weight (water to cement ratio for a standard mix) should be 0.65.*

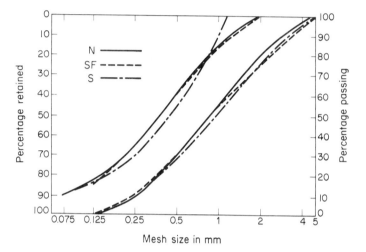

Figure B.2 □ *(Right) Limits for sieve curves for mortar sands in Scandinavia [B.4].*

B.2 □ Aggregate

The aggregate used for mortar is mostly sand as found in nature, with small adjustments. The sand must be well graded so that different sizes of particles are present, forming a smooth sieve curve between certain limits exemplified in Fig. B.1 (ASTM C144 [B.2]). In Fig. B.2 the limiting gradations prescribed in the codes of three Scandinavian countries are reproduced [B.4]. A sand gradation between the limits will give a dense mass of aggregate, requiring a minimum amount of cementitious material for a given strength. At the same time, creep and shrinkage are minimized, and the mortar is economical in use.

Aggregates should be spherical or rounded, since these shapes produce a type of mortar which is easy to work with. Strength and workability of the mortar are usually better when the mortar sand contains comparatively large particles. If the largest particles are too large, however, they may cause stress concentrations and, at worst, they simply may not be accommodated in the joints. As a general rule, the maximum particle size should not be more than $\frac{1}{3}$ or $\frac{1}{2}$ of the thickness of the joint [B.4].

B.3 □ Cementitious material

There are many cementitious materials used for making mortars.

Masonry lime is produced from calcite or limestone or from magnesium limestone ($CaCO_3$ and $MgCO_3$). When calcium carbonate is heated, it decomposes into calcium oxide (CaO) and carbon dioxide at approximately 900 to 1000°C (1650 to 1830°F). When a proper amount of water is added, the burned and crushed lime (quicklime, CaO) is converted to calcium hydroxide [$Ca(OH)_2$]. This hydrated lime is a fine-grained product which is screened and sometimes further ground to a very fine powder. When water is later added to the slaked lime on the building site, a paste mainly of crystalline calcium hydroxide plus a colloid is formed. As the water evaporates, the paste sets. When the paste is exposed to the air, the preliminary setting is followed by gradual hardening, during which the hydroxyl (OH) is replaced by carbon dioxide from the air, forming calcium carbonate again. Lime mortar depends on the carbon dioxide from the air for its hardening and can therefore not harden in a wet environment.

The processes are only roughly outlined above. Normally magnesium oxide is also present; more details can be found in textbooks on engineering materials ([B.5], [B.6]; see also ASTM C5, C207, C91 [B.2]).

The hydrated lime which is marketed in powder form has a unit weight of approximately 650 kg/m³ (40 lb/ft³) and contains up to 98% magnesium and calcium hydroxide. Quicklime (about 75% calcium oxide and 20% magnesium oxide) must be slaked and aged into lime putty before use. The lime putty weighs about 1300 kg/m³ (80 lb/ft³).

Portland cement is made of a mixture of materials: of calcium oxide, silica, aluminum, and iron oxide plus some impurities. Portland cement is standardized and is described in ASTM specifications C150 and C175 [B.2]. When water is added to Portland cement a plastic paste is obtained. The properties depend upon the amount of cement and water. The paste begins to set within a few hours and sets finally within about 10 hours.

After setting, the paste will continue to harden for a very long period of time. The main hardening, however, takes place in the first month, so that after 1 month about 80 to 90% of the final strength is obtained. Portland cement hardens even in water, and the rate of hardening can be varied within rather wide limits by adjusting the proportions of the different constituents and by different curing processes.

In order to add hydraulic properties (ability to harden in a wet environment) to lime mortar (without magnesium oxide), Portland cement may be added to the mortar. Most mortars used for masonry are cement–lime–sand mortars; the limiting proportions for standardized mortars in the United States are shown in Fig. B.3 (data from ASTM C270 [B.1]). Corresponding data for the Scandinavian countries are shown in Fig. B.4 [B.4].

Portland cement gives a fast setting, high strength mortar, whereas lime give a more workable mortar.

Masonry cement contains Portland cement plus other calcareous material (limestone, etc.) in roughly equal amounts and, in addition, some air-entraining agent (ASTM C91 [B.2], [B.4]); it is made to give a mortar that combines the desirable properties of lime mortar and Portland cement mortar without the need to mix on the building site.

Sometimes glue is added to the lime–cement–sand mortar to make a rapidly setting and strong mortar. The costs are high,

Figure B.3 □ Weight of sand per unit weight of cementitious (lime + cement) material in a mortar as a function of the ratio of cement to lime, ASTM C270-66T [B.1].

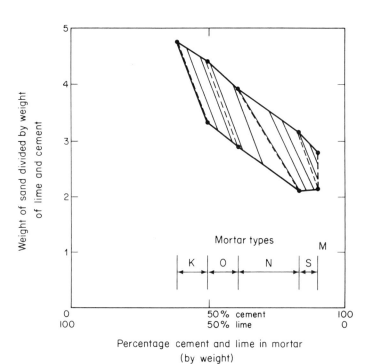

however, and this type of mortar is only used for very thin joints, such as in light weight cellular concrete masonry.

Lately *plastic binders* have been marketed. They add greatly to the tensile strength of masonry, but again the costs are high, and at the present time this type of binder can only be used where it is economically possible or technically necessary.

B.4 ☐ Admixtures

Accelerating additives such as calcium chloride have sometimes been used in mortar. There is a risk of steel corrosion, and staining, however, and the use of calcium chloride is normally not allowed (ASTM C270 [B.1]). Retarding agents are sometimes used to delay the setting of the mortar, for example at building sites with very high temperatures.

Air-entraining admixtures are used to make the mortar more workable and to improve, in some cases, the resistance against freezing. To improve the resistance against early freezing, alcohol is sometimes used. The freezing point can be decreased by 2 to $3\,°C$ by adding 1 to 2% alcohol. Such use is, however, not allowed in the United States (ASTM C270 [B.1]). Water-repelling additives are sometimes used in mortar. These additives are usually wetting agents or silicone powders.

Most of the additives must be added to the mortar in very small amounts, and the recommendations of the manufacturer, the designer, and the codes must be followed strictly. While a correct dose of additive may improve the mortar, it is almost certain that an overdose will harm it. A very strict control over the manufacturing process of such mortar is therefore recommended. The use of admixtures is normally not permitted by American standards (ASTM C270 [B.1]).

B.5 ☐ Classification and some important properties of mortars

The prevailing type of mortar is lime–cement–sand mortar. The amount of different ingredients can vary from pure lime–sand mortar to pure cement–sand mortar. The mortars are usually grouped according to their content of lime and cement. For each group, the maximum and minimum amount of the ingredients are prescribed in codes and specifications. Figure B.4 illustrates the specifications used in the Scandinavian countries. The lime content is calculated on pure 100% calcium hydrate; that is, the

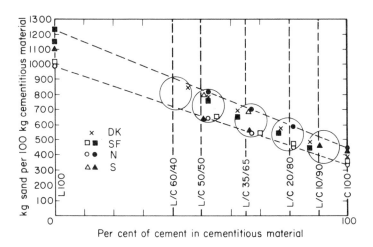

Figure B.4 ☐ *Weight of sand per unit weight of cementitious (pure lime + cement) material in a mortar as a function of the ratio of cement to lime* [*B.4*].

impurities found in practice have been regarded as filler. In Fig. B.3, the ASTM grouping is represented in the same way. Differences in chemical compositions of the "lime" counts for some differences between the curves.

The mason's main requirement for mortar is *workability*. Without good workability, the chances for well-filled mortar joints in the masonry are very low. The workability of a mortar is measured in per cent *flow* according to ASTM Standards C91, C109, C110, and C230 [B.2]. The flow is measured (in the United States) principally as the increase in per cent of the diameter of a molded 2-in. "cake" of fresh mortar which, together with the form bottom but without form sides, is dropped $\frac{1}{2}$ in. 25 times in 15 seconds.

Another important quality for the mortar is the *water retentivity*, which is a measure of the ability of the mortar to retain water and prevent it from escaping into bricks or blocks with high suction. Högberg [B.7] studied the effect of water retentivity on bond and found that a poor water retentivity gives better adhesion to very absorbent materials, but others (Palmer and Parsons [B.8]) claim the opposite. The water retentivity of the mortar is measured in the following way. By vacuum, water is drained from a fresh mortar cake resting on a perforated form bottom covered with filter paper, and the flow is subsequently measured. This value, divided by the flow without suction treatment, is expressed in per cent and called *water retentivity* (ASTM C109, C110 [B.2]).

The *strength* of the (properly proportioned) hardened mortar depends mainly on the cement to lime ratio. Pure lime mortar has a compressive strength of about 1 to 10 kg/cm² (15 to 150 psi), and pure cement mortar has a compressive strength of 150 to 200 kg/cm² (2000 to 3000 psi) or more. The modulus of rupture varies also with the amount of cement added to the lime mortar. Figure B.5 shows one example of the variation of compressive strength and modulus of rupture for the different mortars shown [B.4]. Since other properties also vary with the ratio of lime to cement, the quality of the mortar cannot be judged only by its compressive strength.

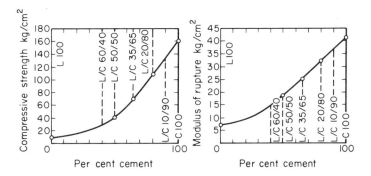

Figure B.5 □ Compressive strength and modulus of rupture of mortars as a function of cement content (only to serve as an example [B.4]).

The *modulus of elasticity* of the mortar plotted as a function of the compressive strength can be seen in Fig. D.7. The *bond* between the mortar and the masonry units is an important property of the masonry, but it is not only the properties of the mortar that determines this. Masonry units with a relatively low suction (less than 20 g/30 sq in. min) and reasonably rough surfaces, and mortar with a reasonably high water retentivity (over 70%) will probably give a good bond. (Compare Fig. K.6).

Sometimes the *frost resistance* is important in a mortar, and sometimes the *water tightness* of the mortar is decisive, especially in areas with heavy, driving rain. *Resistance against chemicals* is also required in some cases. See ASTM Vols. 9 and 12 [B.2], [B.1]. *Shrinkage and creep* will be discussed in Chapter J.

B.6 □ Choice of mortar

When the strength of the masonry is more or less immaterial, a lime–sand mortar with high workability may be chosen. For reasonable strength requirements, a cement–lime–sand mortar of approximately *N*-type (Fig. B.3) may be the best choice. High bending stresses on the masonry will probably be best met with a mortar of *S*-type in combination with units with low suction. In instances where the direct compressive stresses are unusually high, an almost pure cement–sand mortar has to be chosen.

References for Chapter B

B.1 ☐ American Society for Testing and Materials. Part 12: "Chemical-Resistant Nonmetallic Materials; Clay and Concrete Pipe and Tile; Masonry Mortars and Units; Abestos–Cement Products; Natural Building Stones," Book of ASTM Standards, 1967.

B.2 ☐ American Society for Testing and Materials. Part 9: "Cement; Lime; Gypsum," Book of ASTM Standards, 1967.

B.3 ☐ American Society for Testing and Materials. Part 10: "Concrete and Mineral Aggregates," Book of ASTM Standards, 1967.

B.4 ☐ Dührkop, Saretok, Sneck, and Svendsen: "Bruk–Murning–Putsning." (Mortar-Masonry-Plastering.) National Swedish Council for Building Research, Stockholm, 1966.

B.5 ☐ Murphy, Glenn: "Properties of Engineering Materials," 2nd edition. International Textbook Company, Scranton, Pennsylvania, 1947.

B.6 ☐ Neville, Alan M.: "Properties of Concrete." John Wiley & Sons, New York, 1963.

B.7 ☐ Högberg, Erik: "Mortar Bond." National Swedish Institute for Building Research, Report No. 40, Stockholm, 1967.

B.8 ☐ Palmer, L. A., and Parsons, D. A.: "Permeability Tests of 8-in. Brick Wallettes." ASTM Proceedings, Vol. 34, Part II, p. 419, Washington, 1934.

C □ Strength of concentrically loaded masonry walls

C.1 □ Introductory remarks

The most often studied property of masonry, by test and by theory, is its strength under a load perpendicular to the bed joints. This chapter deals with the influences of different factors on the load-carrying capacity of walls loaded in such a manner.

Due to the numerous possible combinations of masonry bricks or units with masonry mortars, the range of obtainable wall strengths is very broad, say 10 to 500 kg/cm² (100 to 7000 psi) [C.1], [C.2], [C.3]. The strength of a masonry wall is also affected by the workmanship, the thickness of the mortar joints, the height of the units, and the age of the mortar, as well as by the suction of the units, as described in the first two chapters of this book.

Within practical limits, the wall strength generally increases with increased brick and mortar strength, so that the brick masonry strength normally is about 25 to 50% of the brick strength, the lower value referring to low strength mortar and the higher value for high strength mortar. The ratio also tends to decrease with increasing brick strength. The concrete masonry strength is normally about 35 to 55% of the block strength (gross area) for some tested types of hollow concrete blocks (Fishburn [C.4], and Richart, Moorman, and Woodworth [C.5]).

The masonry strength is lowered by increasing mortar joint thickness. Tests also show increasing masonry strength with a decreasing ratio of joint thickness to unit height.

A high suction ratio is detrimental to the wall strength—especially in slender columns and in columns subject to eccentric loading, where it seems to affect the evenness of the joint and its modulus of rupture in bending.

C.2 ☐ Effect of masonry unit strength

For a given mortar, the ratio of wall strength to unit compressive strength is often recognized as the "efficiency" of the wall. Factors ranging, at the extremes, from 10 to 90% are reported by Monk [C.1]. For some older test data, the ratios were 10 to 40% for clay masonry, up to 50%, for sand–lime brick masonry and 50 to 90% for concrete brick masonry. These data refer to a variety of column types and slenderness ratios. Tests by Kreüger [C.21] showed an "efficiency factor" of about 20% for brick (25 × 12 × 6.5 cm) masonry prisms 78 to 86 cm (30 in.) high laid in lime mortar with bricks of six different strengths ranging from 135 kg/cm² (1900 psi) to 606 kg/cm² (8500 psi). The "efficiency factor" fluctuated somewhat between 18.5 and 26.5%. Recent tests on single-wythe prism (the prism has the minimum thickness obtainable with the actual brick laid in normal position) test specimens (4 × 8 × 16 in.) with ASTM type *M, S, N,* and *O* mortars gave ratios ranging from 25 to 50% with cored bricks 850 kg/cm² (12,000 psi) [C.6]. There appears to be a slightly better utilization of the brick strength in a single-wythe system than in a multiwythe system.

Considering the strong influence of capping, test machine bearing platens, and the method of testing on brick strength—and also considering the different types of tests for masonry strength—

Notations in Chapter C

A = area of cross section
C = constant
c = core dimension (see Fig. C.7)
d = wall thickness
e = eccentricity of axial load
f' = compressive strength
f_b' = compressive strength of brick
f_m' = compressive strength of masonry
J = moment of inertia
l = wall height
l = length of masonry unit (see Fig. C.7)
P = axial load
S = suction (grams per minute per square decimeter)
s = shell thickness in unit (see Fig. C.7)
IRA = initial rate of absorption (g/30 sq in. min)
σ = stress
σ_u = ultimate compressive stress or strength

---

an "efficiency factor" can only have the value of a rough estimate. Unless the testing equipment is alike, or at least follows certain minimum requirements regarding the stiffness of the heads, and unless the capping of the bricks and wall test specimens is well defined and controlled, comparisons between test results obtained in different laboratories can be misleading. Even under the best of circumstances, a judgment of the wall strength on the basis of the brick strength and an efficiency factor is not well founded theoretically, because it is dubious that failure is governed by the same factors in the two types of loading in all ranges of stresses. Walls usually show vertical cracks after failure; brick test specimens often show 45° cracks after failure, because the mode of failure is affected by the loading platens of the testing machine. In the machine the bricks tend to fail in shear, but in a wall the bricks fail in (combined axial compression and) lateral tension caused by the expanding mortar joints.

In spite of these disturbing facts about test results, there does exist a correlation between tested brick strength and masonry strength. This is shown in Fig. C.1a, which is compiled from test data obtained under controlled conditions and with enough specimens to give reliable mean values SCPRF [C.6], [C.7], [C.19]).

Figure C1(a) □ *Effect of compressive strength of bricks on ultimate compressive strength of masonry walls and prisms. (A masonry prism is a small masonry "wall" consisting of a few bricks and mortar joints.)*

├──── *31* ────┤ *equals 31 4-in. prisms tested, built of bricks in the indicated strength range, and having the average masonry strength shown. M, S, N, O are 4-in. brick masonry prisms built with the indicated ASTM mortars [C.9].*
○ *is 4-in. thick 8-ft high brick masonry walls built with S mortar. 8 is 8-in. thick 8-ft high brick masonry walls built with S mortar.*
□ *is 5-in. thick 8-ft high brick masonry walls built with S mortar in running (A) and stack bond (B). (SCPRF [C.6], [C.7], [C.19]).*

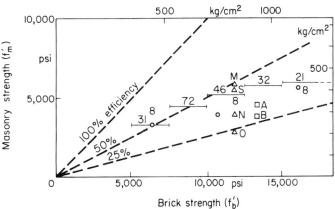

In Fig. C.1b, data from tests reported or made by Nevander [C.20] are shown. In the diagram, best-fit curves of exponential type $[f'_m = C(f'_b)^n]$ are also shown. The lowest "efficiency factor" in Nevander's data is about 5%; the highest is over 50%.

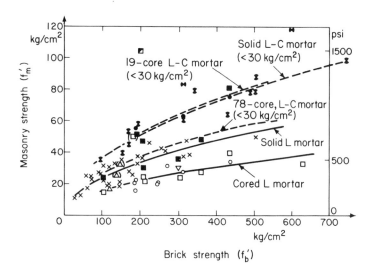

Figure C.1(b) ☐ Relationship between brick strength and masonry strength (Nevander [C.20]).

Mark	Brick	Mortar	
		Type	*Strength (kg/cm²)*
×	Solid	L	
○	19-Core	L	
□	78-Core	L	
▮	Solid	LC	<30
●	19-Core	LC	<30
▢	78-Core	LC	<30
▽	36-Core	L	
△	105-Core	L	
▰	Solid	LC	>30
◪	78-Core	LC	>30

C.3 ☐ Effect of mortar strength

Masonry strength is strongly correlated to the strength of the mortar, a fact which can be clearly seen from Fig. C.2a, plotted from data reported by SCPRF [C.6], [C.7]. The influence of the mortar strength on the masonry strength was also tested by Kreüger [C.21] on piers about 80 cm (30 in.) high, laid in different mortars but of one single brick strength (284 kg/cm² or 4000 psi). The bricks were 25 × 12 × 6.5 cm, and thus the cross section of the piers was about 25 × 12 cm (10 × 5 in.).

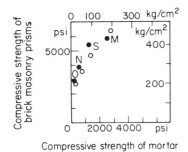

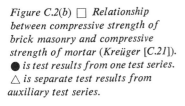

Figure C.2(a) ☐ *Relationship between compressive strength of brick masonry and compressive strength of mortar (SCPRF [C.6], [C.7], [C.19]).*
○ *is 8-in. brick prisms.*
● *is 4-in. walls.*
Brick strength is 830 kg/cm³ (11,800 psi); joint thickness is $\frac{3}{8}$ in. The bricks were wetted when IRA > 20.

Mortars	
Type	Proportions (C : L : S by volume)
M	$1 : \frac{1}{4} : 3$
S	$1 : \frac{1}{2} : 4\frac{1}{2}$
N	$1 : 1 : 6$
O	$1 : 2 : 9$

The results of the tests are shown in Fig. C.2b. A point worth noting is that pure sand mortar (wet sand with no cementitious material) gave the same masonry strength as lime mortar with the compressive strength 2.7 kg/cm² (40 psi). Nylander points out [C.8] that sand-filled joints produce a masonry with a strength about 60% that of masonry with medium strength mortar.

The numerous studies which have attempted to establish a relationship between mortar strength and masonry strength have produced a wide range of results. It appears that masonry strength may vary as the $\frac{1}{3}$ power or the $\frac{2}{3}$ power of mortar strength.

Figure C.2(b) ☐ *Relationship between compressive strength of brick masonry and compressive strength of mortar (Kreüger [C.21]).*
● *is test results from one test series.*
△ *is separate test results from auxiliary test series.*

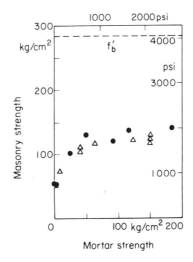

C.4 □ Effect of mortar joint thickness

Since the mortar is usually the weaker part of the masonry composite, the highest strengths are obtained with thin bed joints and a low ratio of bed joint thickness to unit height. For a masonry of approximately 800 kg/cm² (12,000 psi) bricks and approximately 80 kg/cm² (1200 psi) mortar, the relationships between bed joint thickness and masonry brick prism strength is as shown in Fig. C.3.

Kreüger [C.21] points out that, in his experience, the thinner the joints are, the less influence on the strength of the masonry the strength of the mortar has. This observation is in line with the implications of the calculations made by Vinberg [C. 22], which show that a low ratio of joint thickness to brick thickness gives lower lateral stresses in the bricks due to the mortar expansion during loading. The same effect of the joint to brick thickness ratio has been reported by Levicki [C.23].

It seems reasonable to assume that the masonry strength is *decreased* by about 15% for every $\frac{1}{8}$-in. *increase* in bed joint thickness, and vice versa, the normal value being at a normal joint thickness of $\frac{3}{8}$ in. This relationship is, of course, only to be used as a guide for estimating the variation within normal practical limits.

Figure C.3 □ Relationship between brick masonry prism strength and mortar joint thickness. Brick strength is 11,800 psi. Mortar is type S (SCPFR [C.6], [C.7]).
● is 4-in. masonry prisms.
○ is 8-in. masonry prisms.

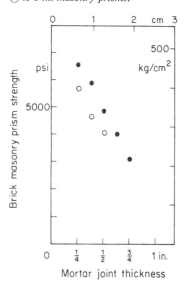

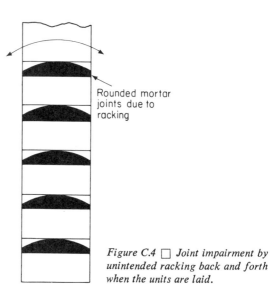

Figure C.4 □ Joint impairment by unintended racking back and forth when the units are laid.

C.5 □ Effect of initial rate of absorption (IRA) (suction)

Haller [C.10] found for one specific case that, for eccentrically loaded walls with the load acting at the kern boundary ($e/d = \frac{1}{6}$, $h/d = 25$), the strength σ of a wall decreased with increasing IRA according to the following equation:

$$\sigma_u = 400/S \text{ kg/cm}^2 \qquad \text{for } 7 < S < 60 \qquad \text{(C.1)}$$

or, in United States units,

$$f' = 11{,}400/\text{IRA psi} \qquad \text{for } 15 < \text{IRA} < 120 \qquad \text{(C.2)}$$

(See Section A.4 and [C.9] for definitions.) The same tendency, but less pronounced, was found for concentrically loaded specimens.

There are indications that the strength decreases again for lower values of IRA, but that a completely wet brick, with IRA = 0 due to the wetting, still gives good strength but has the drawback of long drying time. The phenomenon seems partly to depend on a rapid reduction of the mortar plasticity when it comes into contact with a high suction brick, thereby causing a ridging of the mortar joint due to inevitable small disturbances of the masonry while laying the brick. (See Fig. C.4.)

The effect is more marked for a cement–sand mortar than for a cement–lime–sand mortar (Fig. C.5) according to Haller [C.10]. The phenomenon is more detrimental for eccentric loading than for concentric loading, as can be seen from Fig. C.6, which shows the decrease in strength ratio, eccentric to concentric loading, with increasing suction.

Brick masonry walls tested by Albrecht and Schneider [C.24] show little difference in compressive strengths between masonry

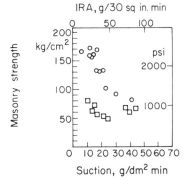

Figure C.5 □ Relationship between strength and brick suction (IRA) for concentrically loaded brick masonry walls (Haller [C.10]).
○ *is cement-sand mortar.*
□ *is cement-lime-sand mortar.*
Wall height to thickness ratio is 21.5.
Wall thickness is 15 cm (≈ 6 in.).

Figure C.6 □ Relationship between P_1/P_0 and suction (IRA) for brick masonry walls (Haller [C.10]).
P_1 is eccentric loading at the kern boundary.
P_0 is concentric loading.
○ *is cement-sand mortar.*
□ *is cement-lime-sand mortar.*

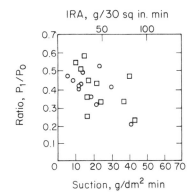

walls laid of low suction cored bricks and those laid of high suction cored bricks, when the difference in brick strength is taken into account. The difference was greater for eccentric loading, and prewetting of the bricks increased the strength of the masonry by a considerable amount for the high suction type. The modulus of elasticity also increased considerably due to the 1-minute wetting of the bricks.

C.6 □ Effect of coring

The influence of cored bricks on the compressive strength of masonry is not fully understood. In some instances the compressive strength of cored units has been higher than that of uncored units. Monk's data [C.1], obtained from tests on masonry prisms of cored and uncored bricks, are summarized in Table C.1. All values are calculated as strength *ratios*, with the compressive strength of the uncored brick 700 kg/cm², (10,090 psi) taken as unity.

The low strength of five-slot masonry prisms is probably due to the notch effect at the corners of the quasi-rectangular cores. The notch effect from the vertical cores is detrimental to the vertical strength of the masonry, since the mortar expands horizontally more than the bricks do, and thus produces a triaxial state of stress with lateral tension in the bricks. The laterally stressed bricks become sensitive to the notches in the vertical cores.

Richart, Moorman, and Woodworth [C.5] report results from tests on walls of cored masonry concrete blocks with different block strengths (40 to 110 kg/cm², 550 to 1550 psi) and different corings (30 to 57%). The ratio of wall to block strength was close to 0.53 in all tests.

Table C.1 □ Effect of coring on compressive strength of clay brick walls [C.1].

Type of coring	Brick				Masonry prism	
	Compressive strength		Modulus of rupture		Compressive strength	
	Gross	Net	Gross	Net	Gross	Net
None	1.00	1.00	0.17	0.17	0.55	0.55
3-hole	1.16	1.35	0.12	0.19	0.47	0.55
5-slot	1.01	1.28	0.083	0.22	0.29	0.37
10-hole	0.89	1.08	0.078	0.14	0.44	0.54

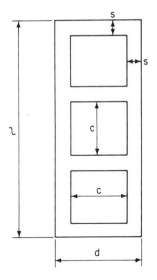

Figure C.7 □ Assumed shape of cross section of a cored masonry unit for study of the effect of coring on the decrease of load-carrying capacity caused by the eccentricity of an axial load.

Nevander [C.20] reports results from tests on masonry walls laid in lime mortar and in lime–cement mortar using bricks that are uncored, cored with 19 holes, and cored with 78 holes (Fig. C.1). The results of these tests show a definite decrease in strength due to the coring for 78-hole bricks and for 19-hole bricks in lime mortar; however, 19-hole bricks in weak [< 30 kg/cm² (425 psi)] lime–cement mortar show the same or higher strength relative to walls of uncored bricks with the same mortar.

In theory, coring should be less detrimental to the strength of eccentrically loaded walls than to the strength of concentrically loaded walls (because of reduction in area), according to the following calculations. Assume that the coring is as shown in Fig. C.7. The following properties of the wall cross section can then be calculated:

$$A = d \cdot l - 3c^2$$

$$J = \frac{ld^3}{12} - 3\frac{c^4}{12} = \frac{ld^3}{12}\left(1 - 3\frac{c^4}{ld^3}\right)$$

The maximum edge stress σ is

$$\sigma = -\frac{P}{dl - 3c^2} - \frac{Pe}{\frac{ld^3}{12}\left(1 - 3\frac{c^4}{ld^3}\right)}\frac{d}{2} \le f'$$

where e is the eccentricity of the axial load. Thus, the maximum load the wall can carry before the ultimate stress f' is exceeded is

$$P = \frac{f'}{\frac{1}{dl - 3c^2} + \frac{6e}{ld^2\left(1 - 3\frac{c^4}{ld^3}\right)}}$$

With $c = \alpha d$, $s = 0.5(d - c)$, and $l = 3c + 4s$, the equation can be rewritten

$$P = \frac{f'd^2(2 + \alpha - 3\alpha^2)}{1 + 6\frac{e}{d}\left(\frac{2 + \alpha - 3\alpha^2}{2 + \alpha - 3\alpha^4}\right)}$$

From the example above, valid for blocks with 3 square cores, $\alpha = 0.9$, and $e = d/6$, the following conclusion can be obtained. The ratio of P for a wall loaded at the kern boundary to P for a concentrically loaded wall is 0.67 for a cored wall, but 0.5 for a solid wall. This value (0.67) would change for other types of coring, but the general effect should be the same.

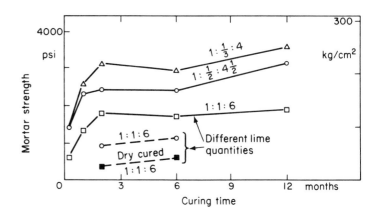

Figure C.8 □ Relationship between mortar strength and curing time, 60°F and 65% relative humidity. Mortars of different lime qualities and mortars of different mix (Davis, [C.11]).

C.7 □ Effect of aging

The curing time of the mortar in the joints affects masonry wall strength in the same way as it affects the strength of poured concrete. The increase in mortar strength with time is illustrated in Fig. C.8, based on data from Davis; see [C.11].

Silen [C.12] studied the increase in masonry strength with time and found that brick walls continue to gain strength long after they have been built (Fig. C.9).

C.8 □ Effect of patterns and methods of bonding

Masonry units can be laid in numerous different patterns by overlapping (displacing in the plane of the wall) the units more or less from one course to another. Stretchers and headers can be mixed in different ways, and certain patterns can be repeated in every second, third, or fourth course, and so on. Such differences do not seem to affect the strength of the masonry. If different heights of units are mixed, for example, a course of low units following a course of higher units, etc., it would be safe to assume, for the purpose of calculation, that the whole wall consisted of low units. The higher of joint thickness to unit height ratio of the lower units would be taken into account this way. (Compare Section C.4.)

If a masonry wall is several wythes thick without cavity, the individual wythes are bonded together with bricks (headers) or metal reinforcement bars (ties) between the wythes. These binders will of course be more highly stressed the higher the shear force in the wall is. Tests on walls with metal ties and regular brick bonding [C.7], as well as tests with different con-

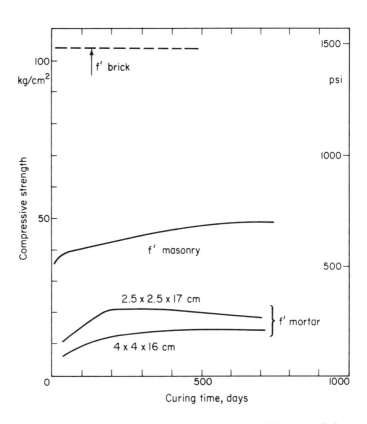

Figure C.9 □ Relationship between hardening time and compressive strength of brick masonry column of lime mortar (Silen [C.12]).

figurations and numbers of binders ([C.1], [C.3]), show little or no difference in wall strength (concentric axial compression) within reasonable sizes of bricks or blocks and with reasonable arrangements of the bonding. This statement holds only if the stress is calculated on the basis of the net area, the vertical joints between the wythes excluded. The unit strength, however, of single-wythe walls seems to be somewhat higher than multi-wythe walls.

C.9 □ The influence of workmanship

Workmanship has a strong influence on the compressive strength of masonry—especially on low strength brick and mortar— according to Monk [C.1], for example. For brick strengths up to approximately 200 kg/cm² (approximately 3000 psi), an increase of up to 100% can be obtained by improving the workmanship over the ordinary. For higher brick strengths Monk reports the gain to be 10 to 70%. The detrimental effect of poor workmanship is due to improper filling of joints and furrowing, as well as to other malpractices which contribute to uneven and incomplete filling of all joints.

C.10 ☐ Influence of variations in dimensions

If the dimensions of the masonry units vary, the dimensions of the mortar joints will also vary. The result is nonuniform joint thicknesses which create bending moments and stress concentrations in the bricks. In order to obtain a high masonry strength, it is advisable to use units of well-controlled dimensions, with a maximum tolerance, say, of the order of a few per cent of the nominal dimensions. In addition, careful control of unit dimensions markedly increases the speed of brick laying, according to Haller [C.13].

C.11 ☐ Strength formulas and the mechanism of failure for masonry subjected to direct compression

A concentrically loaded masonry wall shows considerable vertical cracking before failure. The first cracks appear at a load of roughly half the ultimate load. High strength mortar seems to delay the cracking. At failure the wall exhibits numerous cracks, and is often divided into several separate "columns" at mid height. A typical crack pattern is shown in Fig. C.10, taken from a paper by Albrecht and Schneider [C.25].

The reason for the vertical cracking is fairly well agreed upon. When the masonry is loaded, the bricks and the mortar expand laterally; but since the mortar expands more than the bricks, the bricks are tensioned laterally. The mortar has usually also been "crushed" before the wall fails, and it is prevented from squeezing out of the bed joints by the bricks. This induces a triaxial state of stress in the bricks: vertical compression and bidirectional tension.

Many attempts have been made to formulate the relationship between the strength of the masonry and the unit and mortar strengths. Most of these formulas are, however, not fully based on the fundamental triaxial state of stress in both the bricks and the mortar. They are in some instances simply best-fit formulas developed from test data. Such formulas can nevertheless be useful in practical applications. Recognizing the empirical character of the formulas, some are given here for easy reference.

The following are collected by Ekblad [C.26] and by Graf [C.27]:
Drögsler [C.28]:

$$K_M = 0.736 K_{sb} - \frac{221.5}{k_m} + 28.6$$

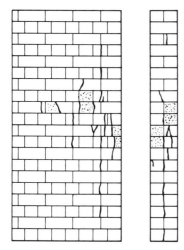

Front side

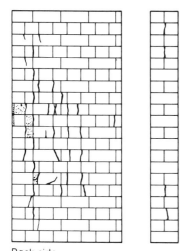

Back side

Figure C.10 □ Typical crack pattern at failure for concentrically loaded brick masonry wall [C.25].

Ekblad [C.26]:

$$K_M = \left(14 + \frac{k_s}{5.9} - \frac{k_s^2}{5200}\right) \frac{\sqrt{k_m} + 0.5}{\sqrt{h} + 12.7} h_s$$

lime mortar, $k_s < 350 \text{ kg/cm}^2$

$$K_M = \left(23 + \frac{k_s}{13}\right) \frac{\sqrt{k_m} + 0.5}{\sqrt{h} + 12.7} h_s$$

lime mortar, $k_s > 350 \text{ kg/cm}^2$

$$K_M = \left(12 + \frac{k_s}{6.5}\right) \frac{\sqrt{k_m}}{\sqrt{h} + 12.7} h_s$$

cement–lime mortar

$$K_M = \left(6 + \frac{k_s}{6.5}\right) \frac{\sqrt{k_m}}{\sqrt{h} + 12.7} h_s$$

cement mortar

Graf [C.29]:

$$K_M = \frac{k_s(4 + 0.1k_m)}{1 + 3h/d} h_s + e \qquad e = 10 \text{ kg/cm}^2 \text{ for excellent work-manship}$$

Haller [C.30]:

$$K_M = (\sqrt{1 + 0.15k_s} - 1)(8 + 0.057k_m)$$

Hansson [C.31]:

$$K_M = 2\sqrt{k_s} + 3\sqrt{k_m}$$

Herrmann [C.32]:

$$K_M = 0.45\sqrt[3]{k_m k_s^2}$$

Kreüger [C.33]:

$$K_M = \left(\frac{k_s}{\gamma^3} + 3.5\gamma^2\right) \frac{6 + 0.1k_m}{8 + 2.5h/d} \sqrt{h_s} \qquad h_s = 6.5 \text{ cm}, \quad k_m \leq 60 \text{ kg/cm}^2$$

Nylander [C.8]:

$$K_M = k_m + k\sqrt[3]{k_s^2} \qquad k = 0.6 \text{ for lime mortar}$$

Oniszczyk [C.34]:

$$K_M = k_s \left(0.33 + \frac{1}{k_s}\right)\left(1 - \frac{0.2}{0.3 + k_m/k_s}\right)$$

Suenson and Dührkop [C.35]:

$$K_M = 3.61\sqrt[4]{k_m} \frac{k_s - 0.15(\max k_s - \min k_s)}{9.71 + \sqrt{k_s}}$$

Voellmy [C.36]:

$$K_M = \frac{k_s}{2} - \frac{180}{\sqrt{k_s}} \left[\frac{\max k_s}{\min k_s} + \left(\frac{k_s - k_m}{100} \right)^2 \right]$$

In these formulas,

K_M = strength of masonry in kg/cm²

k_m = strength of mortar in kg/cm²

k_s = strength of bricks in kg/cm²

k_{sb} = modulus of rupture of bricks in kg/cm²

h_s = height of brick in cm

h = height of wall in cm

d = thickness of wall in cm

γ = density of brick

Still other formulas have been added later on:

Bröcker [C.37]:

$$K_M = \sqrt[3]{k_m} \cdot \sqrt{k_g}$$

and Monk [C.1]:

$f'_m = 0.297 f'_b$ for cement–sand mortar, ratio 1:3

$f'_m = 0.177 f'_b$ for cement–lime–sand mortar, ratio 1:1:6

$f'_m = 0.138 f'_b$ for lime–sand mortar, ratio 0:1:3

$f'_m = -409 + 0.8719 f'_b - 0.3009 (f'_b)^2 \cdot 10^{-4}$ psi

     for S-mortar—SCPRF Laboratory

$f'_m = -1852 + 0.8826 f'_b - 0.2624 (f'_b)^2 \cdot 10^{-4}$ psi

     for S-mortar SCPRF—Commercial Laboratories

In Monk's formulas, the following notations were used: f'_m = strength of masonry; f'_b = strength of bricks.

Hilsdorf [C.14] has, on the basis of extensive measurements on clay bricks in masonry piers, deduced a formula based on the concept of the failure model outlined at the beginning of this section. The theory is based on the assumption that the mortar joints expand laterally under compression more than the bricks do; this subjects the bricks to lateral tension in addition to the externally applied compression at the same time as the mortar is confined, and allows loads to exceed uniaxial strength of mortar. Furthermore, the uneven distribution of the strength properties and thickness of the joints (and the same for the bricks) are taken into account.

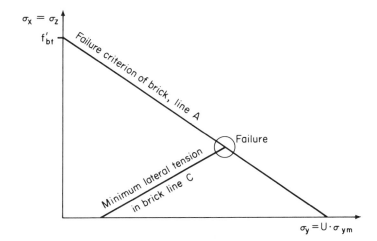

Figure C.11 □ Diagrammatic representation of stresses and failure criteria for bricks and masonry (Hilsdorf [C.16]).

According to Hilsdorf's theory, the factors that affect the compressive strength of the masonry are:

1. The uniaxial compressive strength of the brick.

2. The biaxial tensile strength of the brick.

3. The failure criteria for a brick under a triaxial state of stresses as shown in Fig. C.11, line A. (The external compression and the mortar expansion cause a triaxial state of stress represented by brick cracking at a point somewhere on line A.)

4. The uniaxial compressive strength of the mortar which corresponds to the onset of line C in Fig. C.11.

5. The behavior of the mortar under a state of triaxial compression, determining line C in Fig. C.11.

6. The coefficient of nonuniformity U (a measure of additional stresses from nonuniformity of joints and bricks and their properties).

If line A is assumed to be straight, the following equation describes the failure criterion of the brick.

$$\sigma_x = \sigma_z = f'_{bt}\left(1 - \frac{\sigma_y}{f'_b}\right) \tag{C.3}$$

In which σ_x, σ_y, σ_z are the stresses in the x, y, or z direction; f'_b is the uniaxial compressive strength of brick, and f'_{bt} is the strength of brick under biaxial tension $\sigma_x = \sigma_z$.

The stresses σ_x and σ_z are at the same time the confining stresses on the mortar (line C in Fig. C.11 represents the minimum lateral tensile stress in the brick sufficient to confine the mortar) and if the triaxial strength of the mortar follows the relationship established by Richart, Brandtzaeg, and Brown for concrete [C.15], then the strength of the mortar joint is

$$\sigma_y = f'_j + 4.1\sigma_2 \tag{C.4}$$

where $\sigma_x = \sigma_z = \sigma_2 =$ lateral confining compressive stress in mortar, σ_y is the local compressive stress in mortar, perpendicular to the joint, and f'_j is the uniaxial compressive strength of mortar.

Then the equilibrium conditions for the mean stresses require that

$$\sigma_{xb} \cdot b = \sigma_{xj} \cdot j \tag{C.5}$$

in which σ_{xb} is the lateral tensile stress in bricks, σ_{xj} is the lateral compressive stress in the mortar joint, b is the height of the brick, and j is the thickness of the joint. Substituting σ_{xj} for σ_2 in Eq. (C.4), the equation for line C in Fig. C.11 (stress acting on brick) is obtained:

$$\sigma_x = \frac{j}{4.1b}(\sigma_y - f'_j) = \alpha(\sigma_y - f'_j) \tag{C.6}$$

where $\alpha = j/4.1b$.

The maximum local stress at failure σ_y corresponds to the point of intersection of lines A and C:

$$\sigma_y = f'_b \frac{f'_{bt} + \alpha f'_j}{f'_{bt} + \alpha f'_b} \tag{C.7}$$

Introducing a nonuniformity coefficient at failure U_u, defined as the ratio of maximum stress to average stress over the area, the average masonry stress at failure is calculated by Hilsdorf to be

$$\sigma_{ym} = f'_m = \frac{\sigma_y}{U_u} \tag{C.8}$$

which, when inserted in Eq. (C.7), gives the general expression for the axial compressive strength of masonry:

$$f'_m = \frac{f'_b}{U_u}\left(\frac{f'_{bt} + \alpha f'_j}{f'_{bt} + \alpha f'_b}\right) \tag{C.9}$$

According to Hilsdorf, U_u is a function of quality of workmanship, type and compressive strength of mortar, type of bricks, pattern of masonry, and coring of bricks. U_u is understood to be the value the coefficient has at failure. For lower stresses it can have other values. There was a tendency in Hilsdorf's test for U_u to decrease with increasing mortar strength. The rate of decrease with mortar strength appeared to be different, however, for different mortar types.

The main obstacles which prevent a comparison of Eq. (C.9) with test results are the lack of tests of f'_{bt} and U_u for tests giving values of f'_m. One may also suspect that f'_j in a joint is different from the value of mortar strength obtained from tests on mortar cubes or cylinders, although the difference probably is of less importance in this case.

Tests by Hilsdorf [C.16] gave values of U_u between 1.1 and 2.5 with decreasing values for increasing mortar strength. The writer has, for a crude comparison with other tests, tried two approximative values of U_u:

a. $U_u = 1.5$

b. $U_u = 2 - \dfrac{f'_j}{5000}$ for $f'_j < 4000$ psi

The values of f'_{bt} which are necessary for the comparison are based on the following reasoning. In one reported case the modulus of rupture was observed for three different cored brick types [C.12] and was found to be approximately 10% of the compressive strength. Other tests for many different makes of bricks from all over the United States were compiled by McBurney and Lovewell [C.17]; they show ratios from 0.1 to 0.3 between modulus of rupture and compressive strength of brick. Per unit area, uncored bricks are expected to show about 50% higher modulus of rupture than cored bricks, since the coring reduces the cross section in bending of a brick by approximately one third, and reduces the cross section in compression by about 10% for the type of coring mentioned above [C.12]. Standards in the United States permit up to 25% coring for nonhollow units. Information on the tensile strength of bricks is rare, but McBurney (see paragraph 112 in [C.18]) ran tests on six makes of brick and found the tensile strength to be 30 to 40% of the modulus of rupture.

The information mentioned above gives the following approximate safe ratios of the tensile strength to the compressive strength of bricks: 1:20 for uncored and 1:30 for cored bricks. The ratios are usually higher for low strength bricks and lower for high strength bricks. (If there is any doubt, the tensile strength should be tested for the particular brick and coring.)

For this comparison, two alternative values of f'_{bt} have been chosen:

1. $f'_{bt} = 35 \text{ kg/cm}^2$ (500 psi)

2. $f'_{bt} = \dfrac{f'_b}{30}$ for cored bricks

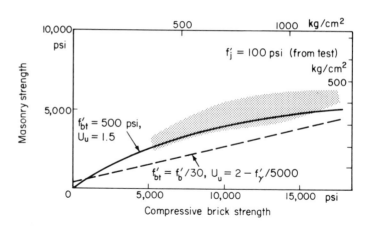

Figure C.12 □ *Relationship between masonry strength and brick strength. Comparison of test data according to Fig. C.1 with theoretical results from Eq. (C.9) with different assumptions for U_u and f'_{bt}.* ▨ *= scatter of test data. f'_j = 1200 psi.*

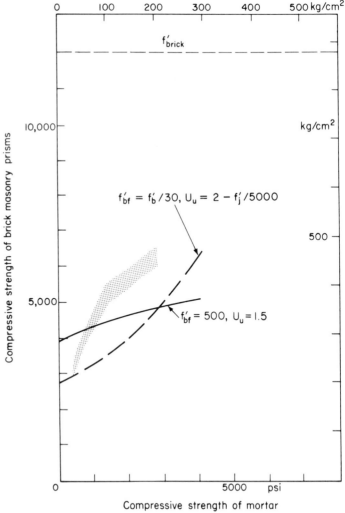

Figure C.13 □ *Relationship between masonry strength and mortar strength. Comparison of test data according to Fig. C.2 with theoretical results from Eq. (C.9) with different assumptions for U_u and f'_{bt}.* ▨ *= scatter of test data. f'_b = 12,000 psi.*

The alternative values of U_u and f'_{bt} discussed earlier have been used only in the combinations a–1 and b–2. The comparison are shown in Figs. C.12, C.13, and C.14.

Taking into account that Hilsdorf's formula was developed mainly from a study on bricks with strengths around 375 kg/cm² (5300 psi), that the present comparison is done on tests for which the tensile strength had to be estimated, that the nonuniformity factor U had to be guessed, and that the bricks for this comparison had high strength, 850 kg/cm² (12,000 psi), the proposed formula seems to be promising; but it should be used with caution until detailed information about U_u and f'_{bt} is obtained for a greater variety of masonry constituents.

Figure C.14 □ *Relationship between masonry strength and mortar joint thickness. Comparison of test data according to Fig C.3 with theoretical results from Eq. (C.9) with different assumptions for U_u and f'_{bt}.*
▨ *= scatter of test data. f'_b = 12,000 psi, f'_j = 1,200 psi.*

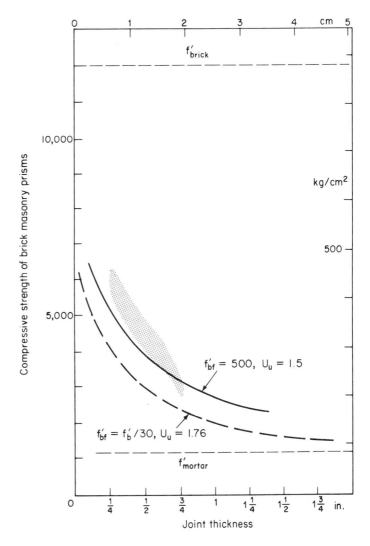

To sum up, the best way to predict masonry strength without pretesting is probably to use one of the more recent empirical formulas based on tests of comparable material (geographical location is probably important). For a study of the influence of a *variation* of geometry, Hilsdorf's formula seems to be best suited, and it could also be used to study the effect of variations in brick strength when the mortar type is unchanged.

References for Chapter C

C.1 ☐ Monk, C. B.: "A Historical Survey and Analysis of the Compressive Strength of Brick Masonry." Research Report No. 12, Structural Clay Products Research Foundation, Geneva, Ill., July, 1967.

C.2 ☐ Stang, Parsons, and Barney: "Compressive Strength of Clay Brick Walls." National Bureau of Standards, Journal of Research, Vol. 3, part 2, 1929.

C.3 ☐ Bragg, J. C.: "Compressive Strength of Large Brick Piers." Bureau of Standards, Technological Papers, No. 111, 1918.

C.4 ☐ Fishburn, C.: "Effect of Mortar Properties on Strength of Masonry." National Bureau of Standards, Monograph 36, Dept. of Commerce. Washington, D. C., November 20, 1961.

C.5 ☐ Richart, Frank E., Moorman, Robert R., and Woodworth, Paul M.: "Strength and Stability of Concrete Masonry Walls." Engineering Experiment Station Bulletin No. 251, University of Illinois, July 5, 1932.

C.6 ☐ Structural Clay Products Research Foundation. "Compressive, Transverse and Racking Strength Tests of Four-inch Brick Walls." Research Report No. 9, Geneva, Ill., 1965.

C.7 ☐ Structural Clay Products Research Foundation. "Compressive and Transverse Strength Tests of Eight-inch Brick Walls." Research Report No. 10, Geneva, Ill., 1966.

C.8 ☐ Nylander, H.: "Undersökning av Bärkraften hos Murade Cementstensväggar." (Investigation of Load-Carrying Capacity of Concrete Block Masonry Walls.) Betong, Häfte 3, Stockholm, 1944.

C.9 ☐ American Society for Testing and Materials, Part 12: "Chemical-Resistant Nonmetallic Materials; Clay and Concrete

Pipe and Tile; Masonry Mortars and Units; Asbestos–Cement Products; Natural Building Stones," 1968.

C.10 ☐ Haller, J. P.: "Die technischen Eigenschaften von Backstein-Mauerwerk für Hochhäuser." Schweizerische Bauzeitung, Heft 28, s. 76, July 12, 1958.

C.11 ☐ Plummer and Blume: "Reinforced Brick Masonry and Lateral Force Design." Structural Clay Products Institute, Washington, D. C., 1953.

C.12 ☐ Silen, H. O.: "Härdningstidens inverkan på med Kalkbruk Murade Konstruktioners bärförmåga." (Influence of the Time of Hardening on the Load-Bearing Capacity of Masonry Structures with Lime-Mortar Joints.) The State Institute for Technical Research. Series III, Building 45, Helsinki, 1961.

C.13 ☐ Haller, J. P.: "Die Entwicklung des Backsteinmauerwerks und die S.I.A.-Norm 113." Schweizerische Bauzeitung 82, Heft 33, s. 579, August 13, 1964.

C.14 ☐ Hilsdorf, H. K.: "An Investigation into the Failure Mechanism of Brick Masonry Loaded in Axial Compression," *Designing, Engineering and Constructing with Masonry Products*, edited by Dr. Franklin Johnson. Copyright ©️ 1969 by Gulf Publishing Company, Houston, Texas. Used by permission.

C.15 ☐ Richart, F. E., Brandtzaeg, A., and Brown, R. L.: "A Study of the Failure of Concrete under Combined Compressive Stresses." Bulletin 185, University of Illinois, Engineering Experiment Station, 1928.

C16 ☐ Hilsdorf, H. K.: "Untersuchungen über die Grundlagen der Mauerwerks-Festigkeit." Bericht No. 40, Materialprüfungsamt für des Bauwesen der Technischen Hochschule, München, 1965.

C.17 ☐ McBurney, J. W., and Lovewell, C. E.: "Strength, Water Absorption and Weather Resistance of Building Bricks Produced in the United States." Proceedings ASTM, Vol. 33, Part II, 1933.

C.18 ☐ Plummer, H., and Reardon, L.: "Principles of Brick Engineering." Structural Clay Products Institute, Washington, D. C., 1939.

C.19 ☐ Structural Clay Products Research Foundation. "Compressive and Transverse Tests of Five-inch Brick Walls." Research Report No. 8, Geneva, Ill., 1965.

C.20 ☐ Nevander, L.E.: "Provningar av Tegelmurverk." (Tests on Brick Masonry.) Tegel No. 5, Stockholm, 1954.

C.21 ☐ Kreüger, Henrik: "Brickwork Tests and Formulas for Calculation." The Clay Worker, London, July and August, (2 Parts) 1917.

C.22 ☐ Vinberg, A. Hans: "Murade Lättbetongväggars hållfasthet." (Compressive Strength of Lightweight Concrete Brick Walls.) Stockholm, 1953

C.23 ☐ Levicki, Bohdan: "Building with Large Prefabricates." (French: "Batiments d'Habitation Préfabriqués en Eléments de Grandes Dimensions.") Warsaw, 1964.

C.24 ☐ Albrecht, W., and Schneider, H.: "Der Einfluss der Saugfähigkeit der Mauerziegel auf die Tragfähigkeit von Mauerwerk," Heft 46, Berichte aus der Bauforschung, Wilhelm Ernst & Sohn, Berlin, 1966.

C.25 ☐ Albrecht, W., and Schneider, H.: "Der Einfluss des Mauerverbandes von 30-cm dicken Hochlochziegel-wänden auf deren Tragfähigkeit." Heft 46, Berichte aus der Bauforschung, Wilhelm Ernst & Sohn, Berlin, 1966.

C.26 ☐ Ekblad, K. G.: "Tegel och murbruk samt murverk av massivtegel." Chalmers Tekniska Högskolas Handlingar, Häfte 84, Göteborg, 1949.

C.27 ☐ Graf, Otto: "Über die Tragfähigkeit von Mauerverk, insbesondere von Stockwerkshohen Wänden." Fortschritte und Forschungen im Bauwesen, Reihe D, Heft 8, Franckh'sche Verlagshandlung, Stuttgart, 1952.

C.28 ☐ Drögsler, O.: "Ziegelbiegefestigkeit und Mauerdruckfestigkeit." Wiener stüd. Prüfanstalt für Baustatik, Folge 1, 1938.

C.29 ☐ Graf, O.: "Versuche mit grossen Mauerpfeilern." Bautechnik, Band 4, Heft 16 and 17, 1926.

C.30 ☐ Haller, P.: "Physik des Backsteins." 1er Teil, Festigkeitseigenschaften, Zürich, 1947.

C,31 ☐ Hansson, O.: "En översikt av Chalmers provningsanstalts murverksprovningar." Häfte 6, Tegel, 1936.

C.32 ☐ Herrmann, M.: "Über die Abhängigkeit der Mauerwerks-festigkeit von der Druckfestigkeit der Steine und des Mörtels unter Berücksichtigung verschiedener Konstruktionseinflüsse. Wissenschaftlicher Abhandl der Deut. Mat. Pruf. Anst., II Folge, Heft 4, 1942.

C.33 ☐ Kreüger, H.: "1942 års normalbestämmelser," Häfte 2, Tegel, Stockholm, 1943.

C.34 ☐ Oniszczyk, in Planen und Bauen, Heft 19, Band 5, 1951, and Polnishe Norm. PN/B182.

C.35 ☐ Suenson, E., and Dührkop, H.: "Forsøg med Murverk af Molersten og almindelig Tegelsten." Ingenørvidenskablige Skrifter, Hefte 1, 1944.

C.36 ☐ Voellmy, A.: "Discussions." International Association for Testing Materials, Congress, London, 1937, p. 423.

C.37 ☐ Bröcker, O.: "Steinfestigkeit und Wandfestigkeit" Betonstein-Zeitung 27, Heft 3, p. 120, 1961.

D ☐ Stability of concentrically loaded masonry walls

D.1 ☐ General considerations

In practice, a masonry wall is often built in or restrained to some extent along the sides. The wall can be continuously built from the bottom to the top and fixed to the floor slabs at every floor level. It can also be rigidly fixed to cross walls, concrete columns, or other structural members. Such restraints are all beneficial to the load-carrying capacity of the wall, especially if the wall is slender. If, on the other hand, the load is acting eccentrically on the wall, or if, in addition to the axial load, wind pressure acts upon it, the load-carrying capacity is lowered. A local reduction in cross-sectional area, for example, by a hole for a window, or by chasings for ducts of different types, is also detrimental to the load-carrying capacity. Poor workmanship and deficient material have the same effect.

It is thus clear that in practice there exists no single-valued load-bearing capacity of a wall as such. To make a meaningful comparison between walls, and to define and test ultimate stresses and stiffnesses of walls, it is necessary to study a masonry wall under well-defined load and boundary conditions. The previous chapter presented studies of basic masonry strength (for example, of a short pier) unaffected by eccentricities, bending moments, buckling phenomena, cut outs, etc. In this chapter we will discuss the strength and stability of a pin-ended masonry wall under concentric load. The "pin-end" boundary condition is often chosen in laboratory tests because it is reasonably simple to achieve, the test is easy to run, and from the evaluation of the test results the basic strength properties of the wall itself are obtained. The complications encountered by an eccentric application of the load will be dealt with in Chapter E, and the interaction between walls and floor slabs will be discussed in Chapter F.

Loading conditions significantly different from the normal, with near vertical inplane axial loading, will be treated later on. Inclined inplane forces will be treated in Chapter G, and significant lateral loadings in Chapter H. The fundamental strength and stiffness factors determining the stability of a masonry wall will be discussed first.

A concentrically loaded pin-ended column or wall buckles at the load

$$P_{cr} = \frac{\pi^2 E I}{h^2} \tag{D.1}$$

where E is the modulus of elasticity, I is the moment of inertia

Notations in Chapter D

A = area of cross section
β = ratio = T_j / T_b
d = thickness of wall
ΔT = compression over the distance T
δ = ratio $T_b / (T_b + T_j) = T_b / T$
E = modulus of elasticity
ε = strain
f' = compressive strength
g = volume of aggregate per unit volume of mix
h = height of wall or column
I = moment of inertia
K = length coefficient (see Fig. D.2)
k = radius of gyration = I/A
l = length of wall
λ = slenderness = h/k
P_{cr} = critical load, the load at which a compressed member fails
σ = stress
T = thickness or joint or brick

Subscripts in Chapter D:

b = brick or block
cr = critical
g = gravel
id = ideal
j = joint
m = masonry
p = paste
u = ultimate

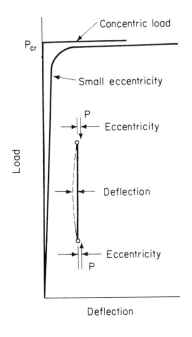

Figure D.1 □ *Relationship between load and deformation for a compressed pin-ended elastic member: concentric load and load with small eccentricity.*

(in the case of a rectangular column, $bd^3/12$), and h is the height of the column between the supports. At this load the column does not straighten if it is bent by a small transverse force which is immediately removed. Due to the inevitable imperfections in the column, the deformations grow beyond all practical limits shortly before the full buckling load is reached, so that the column becomes useless for loads of that magnitude (Fig. D.1). Equation (D.1) can be derived from the differential equation for a centrally loaded column; the details of the derivation are found in most textbooks on strength of materials.

It is obvious from Eq. (D.1) that the load-carrying capacity (the buckling load) of the wall does not depend on the strength of the masonry, but solely on the geometrical dimensions and the modulus of elasticity of the masonry. The buckling *stress* can be calculated from Eq. (D.1) by dividing the buckling load (or, as it may be called, the critical load) by the area of the cross section. Hence, for a rectangular section

$$\sigma_{cr} = \frac{P_{cr}}{A} = \frac{\pi^2 E}{12\left(\dfrac{h}{d}\right)^2} \approx \frac{0.82E}{\left(\dfrac{h}{d}\right)^2} \tag{D.2}$$

Sometimes this equation is rewritten, using the radius of gyration,

$$k = \sqrt{\frac{I}{A}} \tag{D.3}$$

giving

$$\sigma_{cr} = \pi^2 E \left(\frac{k}{h}\right)^2 \tag{D.4}$$

or with the notation $\lambda = h/k =$ the slenderness of the column

$$\sigma_{cr} = \frac{\pi^2 E}{\lambda^2} \tag{D.5}$$

These formulas are different for other end conditions of the wall, as shown in Fig. D.2. For example, the formula for fixed ends is

$$\sigma_{cr} = \frac{4\pi^2 E}{\lambda^2} \tag{D.6}$$

which can be rewritten as

$$\sigma_{cr} = \frac{\pi^2 E}{\lambda_{id}^2} \tag{D.7}$$

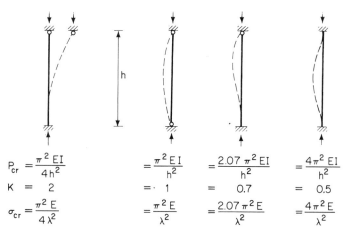

Figure D.2 □ Critical load P_{cr}, length coefficient k, and critical stress σ_{cr} for compressed members with different fundamental end conditions.

$$P_{cr} = \frac{\pi^2 EI}{4h^2} \qquad = \frac{\pi^2 EI}{h^2} \qquad = \frac{2.07\,\pi^2 EI}{h^2} \qquad = \frac{4\pi^2 EI}{h^2}$$

$$K = 2 \qquad\qquad = 1 \qquad\qquad = 0.7 \qquad\qquad = 0.5$$

$$\sigma_{cr} = \frac{\pi^2 E}{4\lambda^2} \qquad = \frac{\pi^2 E}{\lambda^2} \qquad = \frac{2.07\,\pi^2 E}{\lambda^2} \qquad = \frac{4\pi^2 E}{\lambda^2}$$

with the slenderness of the wall now calculated from the expression

$$\lambda_{id} = K\frac{h}{k} \tag{D.8}$$

where K is the length coefficient (in this case 0.5). For one end pinned and one end fixed, $K \approx 0.7$; for one end free and the other end fixed, $K = 2$.

The buckling stress is raised by side restraint of slender masonry walls by cross walls. On the other hand, the benefit of vertical "supports" such as cross walls diminishes with distance from the vertical support, and sometimes a portion of a wall between two window openings works almost as a simple vertical column.

The modulus of elasticity of masonry walls is affected by the properties of the mortar and the brick, as well as by the stress level.

The problem of obtaining a good estimate of E for use in Eqs. (D.1) through (D.8) will be dealt with next.

D.2 □ The modulus of elasticity of masonry walls loaded perpendicularly to the mortar bed joints

D.2a: Theoretical discussion

The modulus of elasticity of masonry is affected by the moduli of both the masonry constituents: mortar and brick. Since it is more or less an accepted fact, supported by test observations, that vertical joints do not affect the modulus of elasticity of the masonry, only the deformations perpendicular to the bed joints will be studied in what follows. If both the mortar and the bricks are assumed to follow Hooke's law, and if it also is supposed that there is no lack of fit at the interfaces, the following relations can be derived.

The total deformation of one brick plus one joint is

$$\Delta T = \Delta T_j + \Delta T_b \qquad (D.9)$$

where ΔT is the total deformation, ΔT_j is the deformation of the joint, and ΔT_b is the deformation of the brick. The deformation ΔT_j of the joint with the thickness T_j is

$$\Delta T_j = \varepsilon_j T_j \qquad (D.10)$$

where ε_j is the strain caused by the stress σ in the joint which has the modulus of elasticity equal to E_j. Using Hooke's law,

$$\sigma = E\varepsilon \qquad (D.11)$$

in Eq. (D.10) gives

$$\Delta T_j = \frac{\sigma}{E_j} T_j \qquad (D.12)$$

Similar equations can be written for ΔT and ΔT_b, and with the modulus of elasticity taken as E_m for the unit consisting of one joint plus one brick, the following equation is obtained:

$$\frac{\sigma}{E_m} T = \frac{\sigma}{E_j} T_j + \frac{\sigma}{E_b} T_b \qquad (D.13)$$

from which the apparent modulus of elasticity E_m can be calculated.

With the thickness ratio $T_j/T_b = \gamma$, and the modulus of elasticity ratio $E_j/E_b = \beta$, Eq. (D.13) can be rewritten as

$$\frac{E_m}{E_b} = \frac{1+\gamma}{1+\gamma/\beta} \qquad (D.14)$$

and if the ratio

$$\frac{T_b}{T_b + T_j} = \frac{T_b}{T} = \delta \qquad (D.15)$$

Eq. (D.13) can be written

$$E_m = \frac{1}{\dfrac{1-\delta}{E_j} + \dfrac{\delta}{E_b}} \qquad (D.16)$$

This equation is similar to the equation

$$E = \frac{1}{\dfrac{1-g}{E_p} + \dfrac{g}{E_g}} \qquad \text{for } E_g > E_p \qquad (D.17)$$

derived by Hansen [D.1] for two-phase material. In Hansen's case (referring to concrete), g is the volume of aggregate per unit volume of mix, E_g is the modulus of elasticity of the aggregate, and E_p is the modulus of elasticity of the cement paste. For $E_g < E_p$, Hansen gives the following formula:

$$E = (1 - g)E_p + gE_g \qquad (D.18)$$

Since Eq. (D.18) is based on a model that does not apply to masonry, only Eq. (D.17) shows similarities with Eq. (D.16). Usually $E_p \lesssim E_g$ $(E_j \leq E_b)$ for masonry.

Theoretically, the modulus of elasticity of the masonry could be calculated from Eq. (D.16), in which E_j could be calculated from Eq. (D.17); i.e.,

$$E_m = \frac{1}{\dfrac{1 - \delta}{E_j} + \dfrac{\delta}{E_b}} \qquad (D.16)$$

with

$$E_j = \frac{1}{\dfrac{1 - g}{E_p} + \dfrac{g}{E_g}} \qquad (D.17)$$

making

$$E_m = f(\delta, E_b, g, E_p, E_g) \qquad (D.19)$$

The application of this equation is limited to the instances where E_p is known (such as for some cement pastes). For lime paste (and porous two-phase systems), data are lacking.

The effect of the various parameters, however, on the modulus of elasticity of masonry can be studied—principally with the aid of Eq. (D.14), which is graphically presented in Fig. D.3. From this figure it can be seen that, for reasonable ratios of joint to brick thicknesses, low ratios of the modulus of elasticity of the joint to the modulus of elasticity of the brick $(E_j : E_b)$ result in low ratios of modulus of elasticity of the masonry to the modulus of elasticity of the brick. It can also be seen that, if the modulus of elasticity of the joint becomes greater than 50% of the modulus of the brick, the ratio of the modulus of elasticity of the masonry to the modulus of elasticity of the brick is affected comparatively little.

Since the literature reports few attempts to relate the modulus of elasticity of the masonry to the modulus of elasticity of the constituents, only a rough check of the validity of Eq. (D.14) can be made. The writer has compared the test results obtained

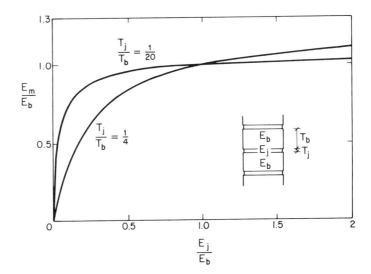

Figure D.3 □ Theoretical relationship between the modulus of elasticity of masonry E_m and the modulus of elasticity of mortar E_j, both expressed in terms of the modulus of elasticity of the bricks E_b: two different joint to brick thicknesses, $T_j/T_b = \frac{1}{20}$ and $T_j/T_b = \frac{1}{4}$.

by Hilsdorf [D.2] and found good agreement with the theory. In Hilsdorf's report, all moduli were taken as the tangent moduli at zero stress; i.e., a tangent to the stress–strain curve was drawn through the origin, and the modulus of elasticity corresponding to this tangent was calculated. Consequently, the agreement between calculated and measured values was good.

For high loads on the masonry—say, for example, loads over 50 kg/cm² (700 psi)—the value of modulus of elasticity for uniaxial loads on lime mortar is meaningless since the crushing strength of such mortar is very low, usually less than 7 kg/cm² (100 psi). For the same loads, the modulus of elasticity of lime mortar confined in a triaxial state of compressive stress similar to that in the mortar joint should be reasonably well predictable from Eq. (D.14). The results reported by Glanville and Barnett [D.3] show that the modulus of elasticity as calculated according to Eq. (D.14) is much higher than the tested values. The discrepancies are probably due to the different levels of stresses at which the modulus was measured. The modulus of elasticity of masonry was measured for the stress range 0 to 350 psi, which was about 0 to 15% of the strength of the cement mortars, but about 0 to 300% of the strength of the lime mortars. These facts must have affected the apparent modulus of elasticity of the masonry, since the measured moduli for at least the lime mortars must have been measured at much lower stresses than for the masonry.

It is thus necessary to estimate the modulus of elasticity for the masonry from moduli of the bricks and mortars at the relevant stress level. This causes a problem, for lime mortars in the masonry are loaded far above their cubic crushing strength. The modulus of elasticity for such mortars should therefore be established under the same confinement conditions as exist in the masonry, in order to give proper results when the test values are used in Eq. (D.14).

Since important data are lacking for some ranges of E_m, it is of interest to study test results of the modulus of elasticity as related to the strength of the material. Such relationships are irrelevant from a theoretical standpoint, but they have practical value. The readily available data regarding the strength of different mortars and bricks are derived from tests which conform to codes; this is not the case with values of modulus of elasticity. In the following sections, test data will be presented and evaluated.

D.2b: Influence of the strength of the masonry units

Glanville and Barnett [D.3] have reported modulus of elasticity for a number of different clay bricks; their results are plotted in Fig. D.4. The data fall reasonably close to a line

$$E_b = 300f' \tag{D.20}$$

Figure D.4 ☐ Relationship between the modulus of elasticity of bricks and the strength of bricks. Result from tests run by Glanville and Barnett [D.3] are marked ● and tests by Hilsdorf [D.2] are marked ☐.

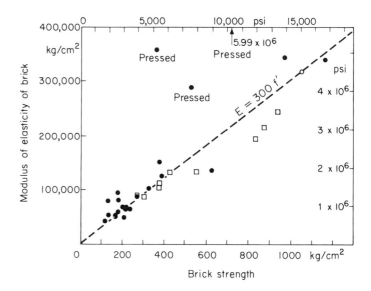

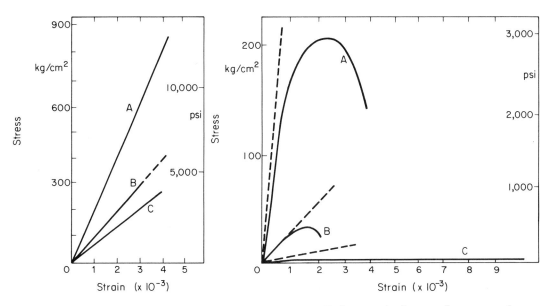

*Figure D.5 ☐ Stress-strain diagrams
for bricks of different strengths.
According to Hilsdorf [D.2],
A has $E_0 = 218,000 \, kg/cm^2 \, (3.1 \cdot 10^6 \, psi)$
B has $E_0 = 126,000 \, kg/cm^2 \, (1.8 \cdot 10^6 \, psi)$
C has $E_0 = 97,000 \, kg/cm^2 \, (1.38 \cdot 10^6 \, psi)$*

*Figure D.6 ☐ Stress-strain diagrams for mortars of
different kinds
A. Cement-sand mortar, ratio 1:3 (by volume)
$E_0 = 253,000 \, kg/cm^2$
$\epsilon_u = 0.31\%$
B. Lime-cement-sand mortar, ratio 1:2:8 (by volume)
$E_0 = 36,000 \, kg/cm^2$
$\epsilon_u = 0.125\%$
C. Lime-sand mortar, ratio 1:3 (by volume)
$E_0 = 6,600 \, kg/cm^2$
$\epsilon_u = 0.88\%$
(Hilsdorf [D.2])*

except for pressed bricks, which have higher modulus of elasticity.

Hilsdorf [D.2] reports data obtained on brick prisms $3 \times 3 \times 6$ cm ($1.2 \times 1.2 \times 24$ in.) cut from different parts of whole bricks (thereby showing the strength distribution over a brick). His data, from three different brick strengths, also fall close to the line represented by Eq. (D.20)—slightly below for high strength bricks. Hilsdorf's measurements showed a good stress-strain linearity for all the tested brick prisms (see Fig. D.5), but Poisson's ratio, about 0.2 at the initial loading stage, increased to about 0.35 before the ultimate load was reached.

Richart, Moorman, and Woodworth [D.4] have reported modulus of elasticity for a number of different types of concrete masonry blocks (Fig. A.10). The results show an approximate relationship between the modulus of elasticity and the block strength:

$$E_u = 1000 f_u' \tag{D.21}$$

or slightly less.

D.2c: Influence of the strength of mortar

Typical stress–strain diagrams, from observations of different types of mortar reported by Hilsdorf [D.2], are shown in Fig. D.6. The tangent moduli of elasticity for zero stress are indicated in the figure. The wide variety of strength and of deformation under loading is striking. The highest strength reported is about 40 times the lowest, and the highest modulus of elasticity reported is about 200 times the lowest. This last figure shows that a careful choice of mortar is necessary when the modulus of elasticity is critical.

The relationship between mortar strength and modulus of elasticity of the mortar should theoretically follow Eq. (D.19). No data on the modulus of elasticity of the lime–cement paste for the low strength mortars are available, a fact which makes it necessary to rely upon observed data for the mortar itself rather than for its constituents.

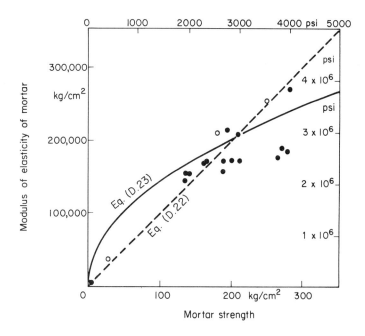

Figure D.7 ☐ Relationship between the modulus of elasticity and the strength of mortar. Results from test run by Glanville and Barnett [D.3] are marked ● and tests by Hilsdorf [D.2] are marked ○. The latter are mean values from three prisms 4 × 4 × 16 cm (or 2.5 × 2.5 × 8 in.) (lime mortar).

Many attempts have been made to fit curves to test data for modulus of elasticity versus strength of concrete—which should behave like mortar. As examples, the following are given:

$$E = 1000f' \tag{D.22}$$

$$E = 33w^{1.5}\sqrt{f'} \text{ psi} \tag{D.23}$$

where w is the unit weight (lb/ft³) of the hardened concrete.

$$E = 1.8 \times 10^6 + 500f' \text{ psi} \tag{D.24}$$

None of these equations is based on physical reasoning; they are all simple "best-fit" curves obtained from tests in the concrete strength range of 150 to 450 kg/cm² (2000 to 6000 psi), which explains the large discrepancy for low strength mortars.

In Fig. D.7 the writer has compiled data reported by Glanville and Barnett [D.3] and by Hilsdorf [D.2]. The expression in Eq. (D.22) is the simplest, and it is close to the observed data for the whole range of different mortars; however, it gives values too low for low strength mortars and too high for high strength mortars. The expression in Eq. (D.23) is better than Eq. (D.22) for high strength mortars. However, it gives calculated values that are too high for low strength mortars.

D.2d : Influence of the strength of masonry

The mortar and the bricks combine to form a masonry whose modulus of elasticity is theoretically affected by both the modulus of elasticity of the mortar and that of the bricks, according to Eq. (D.14) or Eq. (D.19).

Figure D.8 shows stress–strain data given by Hilsdorf [D.2] for cement–sand mortar, for bricks, and for masonry built with these components. In Fig. D.9, corresponding data have been plotted for the same type of bricks, but with low strength lime–sand mortar. The qualitative validity of Eq. (D.19) is obvious; but at the same time the problem of establishing a proper value of the modulus of elasticity for low strength mortars, loaded far above their strength (as measured in standard test cubes or prisms), is clearly to be seen.

In Fig. D.10, tests on brick masonry walls and prisms of different strengths have been plotted in order to give an estimate of the modulus of elasticity as a function of the strength of masonry. The test results were reported by Glanville and Barnett [D.3], Hilsdorf [D.2], and SCPRF [D.5]. Most of the results fall between the values $E = 1000f'_m$ and $E = 400f'_m$.

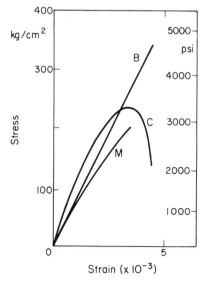

Figure D.8 ☐ Relationship between stress and strain for bricks, for cement-sand mortar, ratio 1 : 3, and for masonry made of these constituents (Hilsdorf [D.2]).
C is cement-sand mortar, ratio 1 : 3
B is bricks
M is masonry

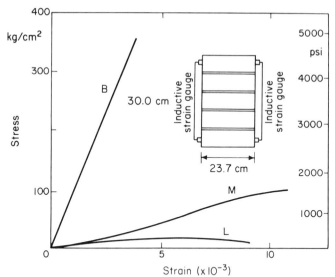

Figure D.9 ☐ Relationship between stress and strain for bricks, for lime-sand mortar, and for masonry made of these components (Hilsdorf [D.2]).
L is lime-sand mortar
B is brick
M is masonry

Figure D.10 ☐ Relationship between the modulus of elasticity and the strength of brick masonry. Test results by Glanville and Barnett [D.3] are marked ●, results by Hilsdorf [D.2] are marked ○, and results by SCPRF [D.5] are marked △.

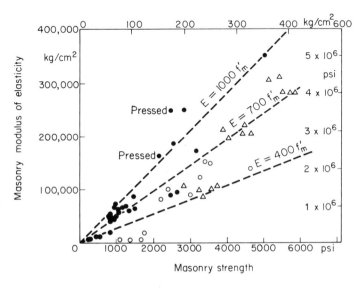

For a rough estimate of the modulus of elasticity of masonry in compression, the following equation can be used:

$$E = 700f'_m \tag{D.25}$$

This equation should be used with caution for very low mortar strength and for unusual ratios of mortar to brick strengths. It should be remembered that a specific masonry strength can be obtained, for example, by combining medium strength mortar with medium strength bricks, or by combining low strength mortar with high strength bricks. It is not to be taken for granted that the modulus of elasticity would be the same in both cases.

In Fig. D.11, tests on concrete masonry walls of different strength and aggregate types have been plotted. The test data were reported by Richart, Moorman, and Woodworth [D.4], and by Fishburn [D.6]. For a rough estimate of the initial tangent modulus (modulus of elasticity for near zero stress), the following equation, somewhat on the low side, can be used.

$$E = 1000f'_m \tag{D.26}$$

The modulus of elasticity determined by Eqs. (D.25) and (D.26) refers usually to low stresses, and with increasing stress the tan-

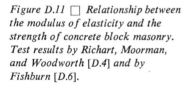

Figure D.11 □ Relationship between the modulus of elasticity and the strength of concrete block masonry. Test results by Richart, Moorman, and Woodworth [D.4] and by Fishburn [D.6].

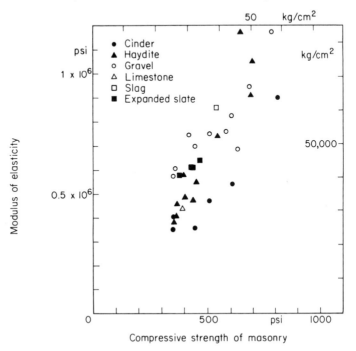

gent modulus normally decreases. In some instances, however, (weak mortar combined with strong units) the strain–stress diagram is S-shaped because the modulus of elasticity first increases and then decreases (Fig. D.9). The S-shape of the stress–strain diagram depends on the strength properties of the mortar and the units. A weak mortar has a low modulus of elasticity to start with; this gives the masonry a low modulus of elasticity near the origin of the stress–strain diagram. When the mortar breaks down, it becomes compacted and confined by the bricks, so that it has a higher modulus of elasticity and a steeper stress–strain curve. Finally, when the bricks start to fail, the modulus of elasticity again decreases and the stress–strain diagram reaches a maximum at the ultimate stress. The influence of the stress level on the modulus of elasticity and the buckling load of masonry walls is studied below.

D.2e : Influence of the stress level

The values of the modulus of elasticity for masonry mentioned above mainly refer to low stresses. With increasing stress, the (tangent) modulus of elasticity will change value. The manner of changes depends upon the type of the mortar. The relationship between the modulus of elasticity for masonry and the stress in the masonry has been derived from the stress–strain curves which Hilsdorf [D.2] reported for different types of mortar, but constant quality of bricks. Figure D.12 shows how materials with different elastoplastic behavior combine into masonry of widely different behavior, even in a case where one constituent (the bricks) is kept constant. The figure has been redrawn in dimensionless form in Fig. D.13.

Cement–sand mortar and lime–cement–sand mortar both give a masonry with decreasing modulus of elasticity under increasing stress. The lime mortar gives a masonry with first an increasing and then a decreasing modulus of elasticity with increasing stress.

Since the tangent modulus varies with the stress, the buckling loads of a masonry column, as well as the deformations under combined bending and thrust, will depend upon this variation. To account for this phenomenon, it is desirable to correlate test data on the modulus of elasticity and the buckling strength of columns made of the same kind of masonry. However, neither the modulus of elasticity obtained from direct compression tests, nor the modulus of elasticity obtained from direct bending tests, applies without reservation to cases with combined loadings

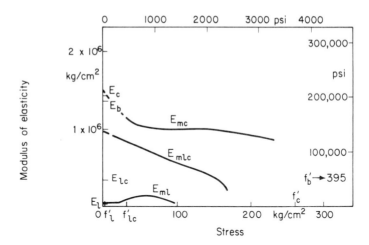

Figure D.12 ☐ *The tangent modulus of elasticity of brick masonry as a function of the applied stress. Curves derived from data given by Hilsdorf [D.2] (30-cm gauge length).*

E_b = *modulus of elasticity of the bricks*
E_c = *modulus of elasticity of cement mortar*
E_{lc} = *modulus of elasticity of lime-cement mortar*
E_l = *modulus of elasticity of lime mortar*
f_b' = *strength of bricks*
f_c' = *strength of cement mortar*
f_{lc}' = *strength of lime-cement mortar*
f_l' = *strength of lime mortar*
E_{mc} = *modulus of elasticity of masonry of cement mortar*
E_{mlc} = *modulus of elasticity of masonry of lime-cement mortar*
E_{ml} = *modulus of elasticity of masonry of lime mortar*

(such as eccentric axial force), since all these moduli can be different. The variations are caused by the uneven distribution of thickness, stress, and strength over the joint. Nylander [D.8] points out the fallacy of the assumption that the stress–strain diagram obtained from a concentrically loaded prism can be used in calculations for a situation which combines bending and axial loading (such as is the case with eccentrically loaded columns.) Nonetheless, the principal effect on buckling of a decrease in modulus of elasticity with increasing stress will, however, be discussed in Section D.3 under the assumptions that the moduli are in fact the same, and that the tangent modulus formula can be applied.

The decrease of the modulus of elasticity with increasing stress was shown for different mortars in Fig. D.13. No buckling tests

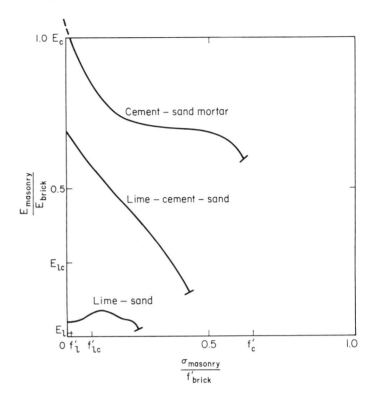

Figure D.13 □ The tangent modulus of elasticity of brick masonry as a function of the applied stress (Fig. D.12 redrawn in dimensionless form).

were run in the same series and since it is desirable to relate observations by maintaining as many constant factors as possible, an examination of the effect on the buckling formula of the relationship between the modulus of elasticity and the stress will be made with the use of another series of buckling tests and prism tests. These calculations lack numerical generality and will serve only as illustrations.

From the stress–strain plots reported by SCPRF [D.5 and D.7], the tangent modulus for increasing values of stress has been estimated and plotted in Fig. D.14. (These plots also show clearly that the modulus of elasticity decreases as the stress increases.) Without access to the primary data, and because of the rather steep (in the chosen scale) stress–strain diagram, the precision of the evaluation must be taken to be relatively low, and the representation has been limited to a few cases of short columns or prisms.

An approximate relationship

$$E_t = 2.75 \cdot 10^6 - 350\sigma \text{ psi} \qquad \text{for } \sigma \lesssim 4000 \text{ psi} \qquad \text{(D.27)}$$

(which fits the data in Fig. D.14 reasonably well up to approxi-

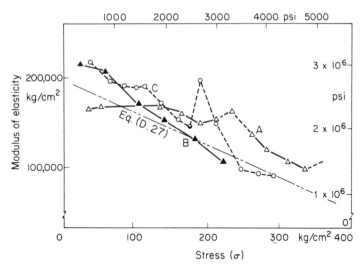

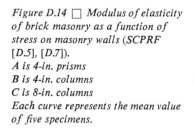

Figure D.14 ☐ Modulus of elasticity
of brick masonry as a function of
stress on masonry walls (SCPRF
[D.5], [D.7]).
A is 4-in. prisms
B is 4-in. columns
C is 8-in. columns
Each curve represents the mean value
of five specimens.

mately 4000 psi) could be chosen for this particular kind of
masonry. The average value of $2.75 \cdot 10^6$ psi for the tangent
modulus at zero stress was reported for several columns. The
application of Eq. (D.27) on buckling of columns will be shown
below. Equation (D.27) is valid only for this test series; it will
be used only for evaluation of test data presented in Section D.3.

D.3 ☐ Buckling stresses of masonry columns : calculations vs. test data

Figure D.15 plots the results from tests on 4-in. and 8-in. brick
masonry walls reported by SCPRF [D.5 and D.7]. The critical
stress is plotted against the height over thickness ratio h/d. The
tests were run with the wall ends fixed against the loading plates
of the testing machine in a manner that prevented rotation of
the ends. This loading case corresponds to a length coefficient of
0.5 in Eqs. (D.7) and (D.8).

If the value of E is taken according to Eq. (D.27), the following equation is obtained:

$$\sigma_{cr} = \frac{\pi^2(2.75 \cdot 10^6 - 350\sigma_{cr})}{\left(0.5\dfrac{h}{k}\right)^2} \tag{D.28}$$

Substitution of $k = \sqrt{I/A} = d/\sqrt{12}$ and solving for σ_{cr} gives

$$\sigma_{cr} = \frac{9 \cdot 10^6}{1150 + \left(\dfrac{h}{k}\right)^2} \text{ psi} \tag{D.29}$$

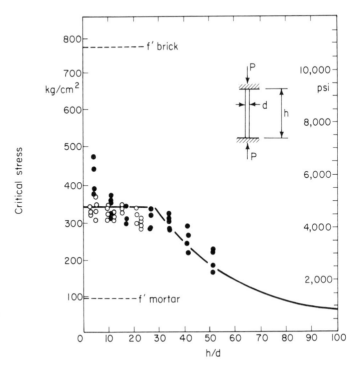

Figure D.15 □ Relationship between the ultimate stress and height over thickness ratio h/d for masonry columns tested by SCPRF [D.5], [D.7]. Theoretical curve is from Eq. (D.31).
● *is 4-in. walls*
○ *is 8-in. walls*

The theoretical values fit the test data reasonably well up to stresses of about 4500 psi, or h/d ratios of approximately 30 (which correspond to $h/d = 15$ for pin-ended columns). It should be observed, however, that no test data are available from this test program for $h/d > 50$ (or stresses below $\sim$ 3000 psi).

For h/d ratios under approximately 30, the mode of failure changes and becomes a stress problem. A compressive strength of approximately 4750 psi (340 kg/cm²) could be taken as a reasonable value for the strength in this specific case. The higher values observed for very small values of $h/d(<5)$ are probably due to the confining effect of loading platens in the test machines. (The ratio h/L was $\sim\frac{2}{3}$.)

The plotted test data are lower than the values predicted by Eq. (D.29), particularly at the intersection of the theoretical buckling curve and the horizontal line drawn at 4750 psi (the line which represents the material strength). This is probably a result of unaccounted for eccentricities of unknown magnitude. Such eccentricities tend to be larger for larger h/d values, and consequently they cause more severe reduction of the load-carrying capacity at higher h/d ratios. For high h/d ratios, the additional eccentricities due to the bending of the wall cause an even greater reduction in strength close to the transition points. Furthermore, it is probable that, close to the crushing strength of the material, the modulus of elasticity decreases more than is assumed in Eq. (D.29). The results of some test series infer that the modulus of elasticity in bending can be different from the modulus of elasticity in direct compression.

Several other approaches to the calculation of the buckling load of a masonry wall can be found in the literature. Haller [D.9], [D.10] assumes a sine-shaped deflection of the wall and uses a stress–strain relationship obtain from concentrically loaded columns. From these assumptions and the equations of epuilibrium at mid height of the wall, and by plotting the axial load as function of a series of assumed edge strains, a maximum load (the buckling load) is obtained. The edge strain that corresponds to the buckling also gives the stress distribution and the deflection at the maximum load.

Angervo [D.11] uses an analytical function of a special shape for the stress–strain relationship, a funchon which enables him to solve the differential equation for an eccentrically (or concentrically) loaded column without tensile strength. The stress–strain function which he assumes can to a certain extent be fitted to curves obtained from tests on concentrically loaded piers.

To summarize, the critical stress of masonry walls with different slendernesses can be calculated with reasonable accuracy from the usual theories of strength of materials, provided that (1) the decresase of the modulus of elasticity with increasing stress, (2) the crushing strength of the masonry, and (3) the load eccentricities are known or can be estimated with reasonable accuracy.

References for Chapter D

D.1 ☐ Hansen, Torben C.: "Creep of Concrete." Swedish Cement and Concrete Research Institute, Bulletin 33, pp. 24–33, Stockholm, 1958.

D.2 ☐ Hilsdorf, H. K.: "Untersuchungen über die Grundlagen der Mauerwerksfestigkeit." Bericht Nr. 40, Materialprüfungsamt für das Bauwesen der Technischen Hochschule, München, 1965.

D.3 ☐ Glanville and Barnett: "Mechanical Properties of Bricks and Brickwork Masonry." Department of Scientific and Industrial Research, Building Research, Special Report No.

22, Building Research Station, Garston, Watford, Herts. Her Majesty's Stationery Office, London, 1934.

D.4 ☐ Richart, Moorman, and Woodworth: "Strength and Stability of Concrete Masonry Walls." Bulletin No. 251, Engineering Experiment Station, University of Illinois, Urbana, Illinois, 1932.

D.5 ☐ Structural Clay Products Research Foundation: Compressive and Transverse Strength Tests of Four-inch Brick Walls." Research Report No. 9, Geneva, Illinois, 1965.

D.6 ☐ Fishburn, Cyrus: "Effect of Mortar Properties on Strength of Masonry." National Bureau of Standards, Monograph 36, 1961.

D.7 ☐ Structural Clay Products Research Foundation: "Compressive and Transverse Strength Tests of Eight-inch Brick Walls." Research Report No. 10, Geneva, Illinois, 1966.

D.8 ☐ Nylander, S. Henrik: "Undersökning av bärkraften hos Murade cementstens-väggar." (Investigation of Load-Carrying Capacity of Cement Block Masonry Walls.) Betong, Häfte 3, Stockholm, 1944.

D.9 ☐ Haller, Paul: "Knickfestigkeit von Mauerwerk aus Künstlichen Steinen." Sweizerische Bauzeitung, Nr. 38 67^e Jahrgang, 1949.

D.10 ☐ Haller, Paul: "Mauerwerk im Ingenieurbau." Schweizerische Bauzeitung, Nr. 18, Februar, 1965.

D.11 ☐ Angervo, Kyösti: "Über die Knickung und Tragfähigkeit eines excentrisch gedrückten Pfeilers ohne Zugfestigkeit." (On the Buckling and the Bearing Capacity of an Eccentrically Compressed Pillar without Tensile Strength.) Staatliche Technische Forschungsanstalt, Finnland; Publication 26, Helsinki, 1954.

E □ Eccentrically loaded walls

E.1 □ Introductory remarks

The fundamentals of the behavior of eccentrically loaded masonry will now be discussed in the light of the general discussion of practical boundary conditions and manner of loading which was presented at the beginning of the previous chapter.

In a direction perpendicular to the bed joints, the tensile strength of a masonry wall is normally very low compared with the compressive strength. In certain types of walls, as for example disc-locked Ytong unit walls (Ytong units are ground light weight cellular concrete blocks with a narrow groove in which small plastic discs are inserted to align the units and lock them together), or the groove and tongue type Siporex block walls, the tensile strength is zero since no mortar or glue is used between the blocks. These circumstances make it desirable to study the structural behavior of a wall made of a material which has no tensile strength when eccentrically loaded (as it often is in a building when it acts together with the floor slabs).

E.2 □ The differential equations for a column which has no tensile strength

E.2a: General

When a column without tensile strength is eccentrically loaded outside the core boundary, at least in some parts of its height, cracks will open and parts of the material will become stressless or "dead" (see Fig. E.1). The extent of the cracked part is determined not only by the end eccentricities of the axial load, but

69

Notations in Chapter E

A = area of cross section

$E_h I_h$ = flexural rigidity of a horizontal structural part (floor slab or horizontal frame member)

$E_v I_v$ = flexural rigidity of a vertical structural part (wall or frame member)

F_1, F_2, F_3 = functions in Eqs. (E.24) to (E.26)

L = length of span of a horizontal structural part (floor slab or horizontal frame member) equals distance between the center lines of vertical structural parts (wall is or vertical frame members)

M = bending moment

P = compressive force

P_c = ultimate load of a vertical structural part (wall) submitted to a central load

b = width of cross section

c = coefficient used in the calculation of the angle of rotation at an end of a wall (see Fig. E.10)

d = thickness of a vertical structural part (wall or vertical frame member)

ε_B = ultimate strain

e = eccentricity of the compressive force acting on a vertical structural part (wall or vertical frame member)

$e_0 = d/6 + pd/3$ = maximum eccentricity of the compressive force, see Eq. (E.10)

f_1, f_2, f_3 = functions in Eqs. (E.14), (E.18), and (E.21), respectively

h = total height of a vertical structural part (wall or vertical frame member) in a story

$m = 6e/d$ = relative eccentricity of the compressive force acting on a vertical structural part

m_1 = relative eccentricity of the compressive force at the top end of a vertical struc-

tural part

m_2 = relative eccentricity of the compressive force at the bottom end of a vertical structural part

p = parameter (cf. e_0)

$s = \sqrt{\dfrac{d/2 - v}{e_0 - v}}$

v = distance from the line of action of the compressive force to the center line of a wall (see Fig. E.2)

x, y, z = coordinates

ε = strain

η = portion of the cross section of a brick masonry wall in compression (see Fig. E.2)

λh = height of a vertical structural part (wall or vertical frame member) from a floor slab to the nearest point of inflection (zero moment point) below this floor slab

σ = stress (compressive stresses are usually reckoned as positive)

σ_c = ultimate stress under the action of a central load

σ_{ult} = ultimate stress; σ_{edge} = edge stress, etc.

φ_h = angle of rotation of an end of a horizontal structural part (floor slab or horizontal frame member)

φ_v = angle of rotation of an end of a vertical structural part (wall or vertical frame member)

φ_{v1} = angle of rotation of the top end of a vertical structural part (wall or vertical frame member)

φ_{v2} = angle of rotation of the bottom end of a vertical structural part (wall or vertical frame member)

$\psi = \tan^{-1}\dfrac{c_1 - c_2}{h} = \text{arc tan }\dfrac{c_1 - c_2}{h}$

also by the deflections and the magnitude of the load. As the load increases, the column deflects and the eccentricity of the force at the central parts of the column (Fig. E.1) increases while the compressed and active section decreases.

The basic differential equation for a column with no tensile strength has been solved by Angervo, 1954 [E.1], and by Chap-

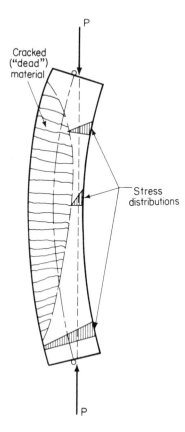

P

Cracked ("dead") material

Stress distributions

P

Figure E.1 □ Schematic illustration of deflection and cracking of an eccentrically loaded elastic column without tensile strength.

man and Slatford 1957 [E.2]. A method of calculation in which the second-order deflections are taken into account in the determination of the angles of rotation at the top and bottom ends of a wall or column has been devised by Angervo, 1957 and 1961 [E.3]. (The moment at mid height of the column shown in Fig. E.1 increases when the load and the deflection increase, thereby accelerating the deflection process.) The equations deduced by Angervo are briefly given in what follows. The loading case studied is illustrated in Fig. E.2.

E.2b: The compressive force lies outside the kern

From geometrical considerations in Fig. E.2, the size of the compressive zone is obtained:

$$\eta = 3\left(\frac{d}{2} - v\right) \tag{E.1}$$

under the assumption that the stress diagram is triangular. With the length of the wall as b, the edge stress is calculated to be

$$\sigma = -\frac{2P}{b\eta} \tag{E.2}$$

and, with the modulus of elasticity taken as E_v, the edge strain is obtained

$$\varepsilon = -\frac{2P}{E_v b\eta} \tag{E.3}$$

Two adjacent cross sections at the coordinates x and $(x + dx)$ are inclined

$$\frac{\varepsilon \, dx}{\eta} = \frac{dv}{dx} \tag{E.4}$$

to each other since the edge is compressed $\varepsilon \, dx$ on the distance dx and the length of the stressless "fiber" at the neutral layer is not affected by the deformation.

By differentiation of Eq. (E.4) we obtain

$$\frac{d^2v}{dx^2} = \frac{\varepsilon}{\eta} \tag{E.5}$$

After inserting Eqs. (E.3) and (E.1) in Eq. (E.5), we obtain

$$\frac{d^2v}{dx^2} = -\frac{2P}{9E_v b\left(\frac{d}{2} - v\right)^2} \tag{E.6}$$

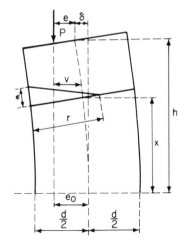

Figure E.2 ☐ Elementary part of a masonry wall in a deflected state, notations (Angervo and Putkonen [E.3]).

By integrating this equation, we get

$$\frac{dv}{dx} = \mp \frac{2}{3} \sqrt{\frac{P}{E_v b}} \sqrt{C - \frac{1}{\frac{d}{2} - v}} \tag{E.7}$$

where C, in view of the condition for dv/dx at $x = 0$, equals $1/[(d/2) - e_0]$. After integrating this equation, we find

$$x = \pm \frac{3}{2} \left(\frac{d}{2} - e_0\right)^{3/2} \sqrt{\frac{E_v b}{P}} \left(\frac{s}{s^2 - 1} + \frac{1}{2} \ln \frac{s+1}{s-1}\right) \tag{E.8}$$

where

$$s = \sqrt{\frac{\frac{d}{2} - v}{e_0 - v}} \tag{E.9}$$

In order to obtain the final formulas in a simpler form, we introduce

$$e_0 = \frac{d}{6} + p\frac{d}{3} \tag{E.10}$$

$$-0.5 \leq p < 1 \tag{E.11}$$

$$m = \frac{6e}{d} \tag{E.12}$$

Then, after inserting $x = h$ and $v = e$, Eqs. (E.7) and (E.8) transform into

$$\left(\frac{dv}{dx}\right)_{v=e} = \mp \frac{d}{3} \sqrt{\frac{P}{E_v I_v}} \sqrt{\frac{1 + 2p - m}{(1 - p)(3 - m)}} \tag{E.13}$$

and

$$h\sqrt{\frac{P}{E_v I_v}} = f_1(p, m)$$

$$= \frac{1}{2} \sqrt{(1 - p)(3 - m)(1 + 2p - m)}$$

$$+ \frac{1}{2} (1 - p)^{3/2}$$

$$\times \ln \frac{2 - m + p + \sqrt{(3 - m)(1 + 2p - m)}}{1 - p} \tag{E.14}$$

which are valid in the intervals

$$\left.\begin{array}{c} 1 \leq m < 3 \\ \frac{m-1}{2} \leq p < 1 \end{array}\right\} \tag{E.15}$$

E.2c: The compressive force lies within the kern at an end of the wall but outside in some parts of the wall

In this case we find the common differential equation

$$\frac{d^2v}{dx^2} = -\frac{Pv}{E_v I_v} \tag{E.16}$$

for the end region. After integrating this equation, and noticing that for $v = d/6$ (i.e., when the line of action of the force coincides with the kern), the value of dv/dx must satisfy Eq. (E.7), we obtain

$$\left(\frac{dv}{dx}\right)_{v=e} = \mp \frac{d}{6} \sqrt{\frac{P}{E_v I_v}} \sqrt{\frac{1+3p}{1-p} - m^2} \tag{E.17}$$

On integrating this equation, and in view of the conditions at the points $v = d/6$ and $v = e$, we find

$$h\sqrt{\frac{P}{E_v I_v}} = f_2(p, m) = \sqrt{p(1-p)} + \frac{1}{2}(1-p)^{3/2}$$

$$\times \ln \frac{1+\sqrt{p}}{1-\sqrt{p}} + \text{arc sin} \sqrt{\frac{1-p}{1+3p}}$$

$$- \text{arc sin } m \sqrt{\frac{1-p}{1+3p}} \tag{E.18}$$

which is valid in the intervals

$$\left.\begin{array}{l} 0 \le m < \text{i} \\ 0 \le p < 1 \end{array}\right\} \tag{E.19}$$

E.2d: The compressive force is situated everywhere within the kern

After integrating Eq. (E.16) and noticing that for $v = e_0$, $dv/dx = 0$, we find

$$\left(\frac{dv}{dx}\right)_{v=e} = \mp \frac{d}{6} \sqrt{\frac{P}{E_v I_v}} \sqrt{(1+2p)^2 - m^2} \tag{E.20}$$

and

$$h\sqrt{\frac{P}{E_v I_v}} = f_3(p, m) = \frac{\pi}{2} - \text{arc sin} \frac{m}{1+2p} \tag{E.21}$$

which are valid in the intervals

$$\left.\begin{array}{l} \dfrac{m-1}{2} \le p < 0 \\ 0 \le m \le 1 \end{array}\right\} \tag{E.22}$$

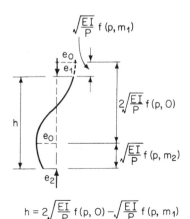

$$\sqrt{\frac{EI}{P}}\,f(p, m_1)$$

$$2\sqrt{\frac{EI}{P}}\,f(p, 0)$$

$$\sqrt{\frac{EI}{P}}\,f(p, m_2)$$

$$h = 2\sqrt{\frac{EI}{P}}\,f(p, 0) - \sqrt{\frac{EI}{P}}\,f(p, m_1) + \sqrt{\frac{EI}{P}}\,f(p, m_2)$$

Figure E.3 □ *Elastic curve of a masonry wall, composed of elementary parts (shown in Fig. E.2 [E.3]).*

E.2e: Combinations of solutions and reference to the undisturbed system

As has been shown by Angervo, the actual shapes of the elastic curve can be represented with the help of combinations obtained by addition and subtraction of the appropriate values of $h\sqrt{P/E_v I_v}$ calculated from the equations above. The method to be used for this purpose is outlined in principle in Fig. E.3. In performing the calculation, the same value of p, that is to say the same value of e_0, must be inserted for all functions.

The angles of rotation at the top and bottom ends of the wall are obtained from Eqs. (E.7), (E.13), and (E.17) by replacing dv/dx, which represents the tangent of this angle, by the angle itself. This angle is measured with reference to the line of action of the compressive force. Therefore, the angle between the line of action of this force and a straight line passing through the centers of the end surfaces of the wall must be added to the angle mentioned above.

If a displacement of the joint between the wall and the slab can occur, then the slope of the wall due to this cause must also be taken into consideration. This slope gives rise to the angle of rotation

$$\psi = \arctan\frac{c_1 - c_2}{h} \approx \frac{c_1 - c_2}{h} \tag{E.23}$$

where c_1 and c_2 are the respective amounts of horizontal displacement at the top and bottom ends of the wall. ψ, c_1, and c_2 are taken to be positive, as shown in Fig. E.4. The subscripts 1 and 2 refer to the top and bottom ends of the wall, respectively.

Finally, the total angle of rotation at the top end of the wall is obtained in the following form (see Fig. E.4, where, as an exception of rotation φ_v is denoted by φ_1):

$$\frac{h}{d}(\varphi_v + \psi) = \frac{m_1 \pm m_2}{6} \pm \frac{h}{p}\left(\frac{dv}{dx}\right)_{v=e_1} \tag{E.24}$$

Figure E.4 □ *Angles of rotation at the top and bottom ends of a story-high masonry wall. In calculating these angles, the horizontal displacements of the joints were also taken into account (Angervo and Putkonen [E.3]).*

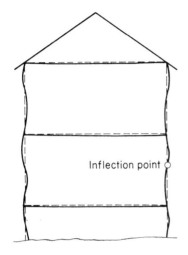

Figure E.5 ☐ Schematic illustration of deflections in a simplified building with load-carrying walls and loaded slabs.

We can distinguish between the three cases dealt with in what follows:

1. The line of action of the compressive force lies outside the kern at the top end of the wall. Then we have

$$\frac{h}{d}(\varphi_v + \psi)$$

$$= \frac{m_1 \pm m_2}{6} \pm h\sqrt{\frac{P}{E_v I_v}}\frac{1}{3}\sqrt{\frac{1 + 2p - m_1}{(1 - p)(3 - m_1)}}$$

$$= \frac{m_1 \pm m_2}{6} \pm h\sqrt{\frac{P}{E_v I_v}}\,F_1(p, m_1) \qquad (E.25)$$

This equation is valid in the same intervals as Eq. (E.14).

2. The line of action of the compressive force lies within, or coincides with, the kern but is situated outside the kern in some section of the wall or its imaginary extension. Then we have

$$\frac{h}{d}(\varphi_v + \psi) = \frac{m_1 \pm m_2}{6} \pm h\sqrt{\frac{P}{E_v I_v}}\frac{1}{6}\sqrt{\frac{1 + 3p}{1 - p} - m_1^2}$$

$$= \frac{m_1 \pm m_2}{6} \pm h\sqrt{\frac{P}{E_v I_v}}\,F_2(p, m_1) \qquad (E.26)$$

This equation is valid in the same intervals as Eq. (E.18).

3. The line of action of the compressive force lies within the kern. Then we obtain

$$\frac{h}{d}(\varphi_v + \psi) = \frac{m_1 \pm m_2}{6} \pm h\sqrt{\frac{P}{E_v I_v}}\frac{1}{6}\sqrt{(1 + 2p)^2 - m_1^2}$$

$$= \frac{m_1 \pm m_2}{6} \pm h\sqrt{\frac{P}{E_v I_v}}\,F_3(p, m_1) \qquad (E.27)$$

This equation is valid in the same intervals as Eq. (E.21).

In the following calculations we assume $\psi = 0$. The functions f_1 to f_3 in Eqs. (E.14), (E.18), and (E.21), together with the functions F_1 to F_3 in Eqs. (E.25), (E.26), and (E.27), give a solution of the differential equation in a parametrical form (p, m).

E.3 ☐ Deformation and maximum edge stresses of the wall

E.3a: The use of the solution to the differential equation

An external wall (see Fig. E.5) can normally be divided into two parts separated at the point of inflection. Each part then consists of a wall centrally loaded at one end (at the inflection point) and

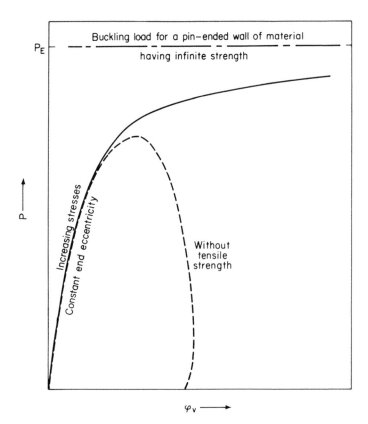

Figure E.6 ☐ *Relationship between*
load and angle of rotation of an
eccentrically loaded wall.
———— *is fully elastic with tensile*
strength.
----- *is elastic in compression*
without tensile strength.

eccentrically loaded at the other end (at the slab end). The load
deformation relationship of the part can—after some calculations
on the basis of Eqs. (E.14), (E.18), (E.21), (E.25), (E.26), and
(E.27)—be represented by the dashed line shown in Fig. E.6 for
one specific eccentricity. For a material having tensile strength,
the curve would be the solid one with an asymptote at P_E. The
maximum edge stress is increasing along the curve. The stresses,
calculated on the basis of a linear stress distribution, are given in
the following equations:

$$\sigma_{\text{edge}} = \frac{4P}{3bd\left(1 - 2\dfrac{e}{d}\right)} \tag{E.28}$$

when the line of action of the compressive force force lies outside the kern and

$$\sigma_{\text{edge}} = \frac{P}{bd}\left(1 + 6\frac{e}{d}\right) \tag{E.29}$$

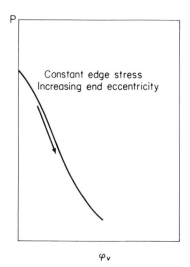

Figure E.7 □ Relationship between load and end rotation for a wall, constant edge stress.

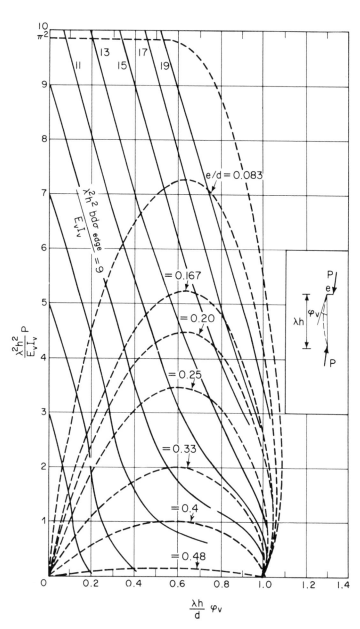

Figure E.8 □ Angles of rotation of, and edge stresses in, a wall having no tensile strength. One end of the wall is subjected to an eccentric load, while the other end is subjected to a concentric load.

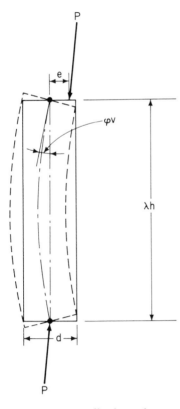

Figure E.9 □ Wall subjected to eccentric compression; the part of the wall between the floor slab and the point of inflection.

Figure E.10 □ Effect produced by the eccentricity of the load on the real flexural rigidity of a masonry wall whose tensile strength is assumed to be equal to zero (Nylander [E.5]).

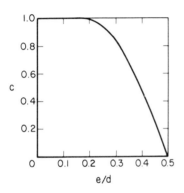

when the line of action lies within the kern. In both of the latter equations the maximum eccentricity e within the wall must be chosen of either e_0, e_1, or e_2.

If on curves of the type shown in Fig. E.6 points of equal edge stress are connected, curves of the type shown in Fig. E.7 are obtained.

In Fig. E.8, curves for constant edge stress and constant end eccentricity according to Sahlin [E.4] are shown. From the curves, the following observations can be made. Contrary to what is the case for a material with tensile strength, the load-deflection curves have maxima for definite deflections φ_v. For low compressive strengths, the edge stress might reach the ultimate stress before the instability load is reached. The failure of the wall can therefore be governed either by elastic instability, with "cracking" on the tensile side, or by crushing on the compressive side.

Thus even for eccentric loads there exists elastic instability, contrary to the case of a linear elastic material which has tensile strength. In the latter case the deformations and stresses grow over all limits before the instability load P_E is reached, and in that case the failure is governed by material failure.

Buckling loads for varying load eccentricities are calculated in Section E.5.

E.3b : Methods of approximation

A method of approximating the angle of rotation at the top or bottom of the wall was proposed by Nylander, 1944 [E.5]. Only the first-order effects due to the cracking in the tensile zone were taken into account and these effects were considered by placing a coefficient c in the well-known formula for the end rotation of a beam loaded by a moment at one end. With $M = Pe$ and the wall height $= \lambda h$, the formula becomes

$$\varphi_v = \frac{Pe \cdot \lambda h}{c3E_vI_v} \tag{E.30}$$

where E_vI_v is the flexural rigidity of the wall and c is obtained from the diagram shown in Figs. E.9 and E.10. This formula gives good results when the load P is much less than the buckling load for the actual eccentricity e. (Compare Fig. E.8.) The location of the zero moment point (inflection point) must also be known or estimated.

78

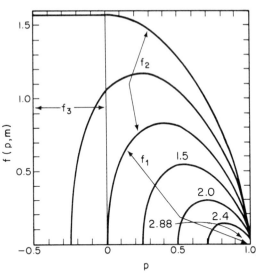

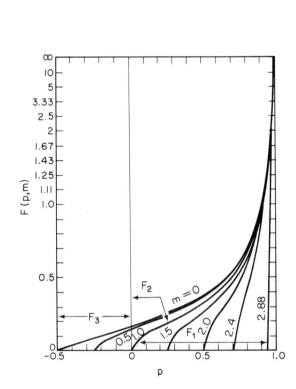

Figure E.11 □ *Graphical representation of the functions F_1 to F_3 and f_1 to f_3, which are used for calculating the angles of rotation of walls having no tensile strength* [E.4].

E.4 □ Diagrams and tables

Since plotting diagrams of the type shown in Fig. E.8 requires use of Eqs. (E.14), (E.18), (E.21), (E.25), (E.26), and (E.27), and therefore involves rather cumbersome calculations, values of the functions f_1 through f_3 and F_1 through F_3 have been calculated by the writer and published in [E.4], from which publication Tables E.1 and E.2 are cited, along with their graphical representation (see Fig. E.11). With the aid of these function values, solutions to the differential equation can be found for any combination of wall parts, as demonstrated in Fig. E.3. The procedure may require some interpolation in the tables.

Table E.1 □ Values of functions f_1, f_2, and f_3 according to Eqs. (E.14), (E.18), and (E.21).

p \ m	0.0	0.5	1.0	1.5	2.0	2.4	2.88
− 0.25	1.5708	0					
− 0.20	1.5708	0.5857					
− 0.15	1.5708	0.7752					
− 0.10	1.5708	0.8957					
− 0.05	1.5708	0.9818					
0	1.5708	1.0472	0				
0.05	1.5689	1.0982	0.4285				
0.1	1.5621	1.1330	0.5796				
0.2	1.5249	1.1635	0.7395				
0.3	1.4708	1.1624	0.8185	0.3167			
0.4	1.3859	1.1217	0.8364	0.4835			
0.5	1.2753	1.0498	0.8116	0.5390	0		
0.6	1.1385	0.9484	0.7509	0.5353	0.2632		
0.7	0.9734	0.8172	0.6571	0.4881	0.2957	0	
0.8	0.7741	0.6525	0.5291	0.4017	0.2655	0.1364	
0.9	0.5227	0.4404	0.3575	0.2735	0.1871	0.1137	
0.99	0.1527	0.1277	0.1025	0.0773	0.0521	0.0319	0.0070

Table E.2 □ Values of functions F_1, F_2, and F_3 according to Eqs. (E.25), (E.26), and (E.27).

p \ m	0.0	0.5	1.0	1.5	2.0	2.4	5.88
− 0.25	0.0833	0					
− 0.20	0.1000	0.0553					
− 0.15	0.1167	0.0817					
− 0.10	0.1333	0.1041					
− 0.05	0.1500	0.1247					
0	0.1667	0.1443	0				
0.05	0.1834	0.1633	0.0765				
0.1	0.2003	0.1822	0.1111				
0.2	0.2357	0.2205	0.1667				
0.3	0.2746	0.2616	0.2182	0.1029			
0.4	0.3191	0.3081	0.2722	0.1925			
0.5	0.3727	0.3632	0.3333	0.2722	0		
0.6	0.4410	0.4330	0.4082	0.3600	0.2357		
0.7	0.5358	0.5292	0.5092	0.4714	0.3849	0	
0.8	0.6872	0.6821	0.6667	0.6383	0.5774	0.4303	
0.9	1.0138	1.0104	1.0000	0.9813	0.9428	0.8607	
0.99	3.3208	3.3198	3.3166	3.3112	3.2998	3.2773	3.0430

E.5 ☐ Buckling loads for varying eccentricities of the axial load

Under the assumption that the material has no tensile strength, but behaves elastically in compression and possesses an ultimate stress σ_B, Angervo [E.1] calculated, from Eqs. (E.14), (E.18), and (E.20), buckling diagrams for walls loaded with an axial load applied with equal eccentricities at both ends. He also calculated buckling curves for walls built of material having curvilinear stress-strain relationships of the following type:

$$\sigma = \frac{\sigma_B}{k_2}\left(1 - \frac{1}{\sqrt{2k_1 \dfrac{\varepsilon}{\varepsilon_B} + 1}}\right) \tag{E.31}$$

where k_1 and k_2 are constants, and σ_B is the ultimate stress of the masonry.

Equation (E.31) can be treated the same way as linear elastic material, but no buckling curves for such material are included here.

In a paper [E.6] eleven buckling diagrams for eccentrically loaded columns or walls made of elastic material with no tensile strength are reported. The degree of eccentricity of the axial load applied was varied in respect to its signs: $+$ and $-$. In Figs. E.12a to E.12e such diagrams for different ratios of eccentricities are reported. Thus $k = 1$ would coincide with the curves for the linear case calculated by Angervo. The appearance of ε_B in the diagrams is due to the fact that the quotient $P/E_v l_v$ in Eqs. (E.14), (E.18), and (E.21) can be rewritten as

$$\frac{12P}{E_v bd^3} = \frac{12\sigma}{E_v d^2} = \varepsilon \frac{12}{d^2}$$

The effect of initial imperfections has been studied by Chapman and Slatford [E.2]. The effect of initial curvature and accidental load eccentricity is similar to the effect of load eccentricity on an initially perfect wall. Thus the effect can be studied, at least in principle, with the aid of Figs. E.12a through E.12e.

The theoretical buckling diagrams in Fig. E.12 are now compared with the test results shown in Fig. D.15. With the ultimate stress still assumed to be 4750 psi (350 kg/cm²), the data on the vertical axis in Fig. E.12a is defined. The mean stress σ_m will be equal to the ultimate stress σ_B at failure (1.0 on the scale). Along the horizontal axis, $\sqrt{\varepsilon_B}$ must be defined for the actual tests.

Figure E.12 a, b, c, d, e □ Buckling curves for eccentrically loaded elastic walls or columns without tensile strength.
The quotient P/E_vI_v in Eqs. (E.14), (E.18), and (E.21) can be rewritten as

$$\frac{P12}{E_vbd^3} = \frac{\sigma 12}{E_vd^2} = \epsilon\frac{12}{d^2}$$

The eccentricities at both ends are indicated in the figures.
  h = wall height
  d = wall thickness
  e = eccentricity
  m = relative eccentricity = 6(e/d)
  σ_B = ultimate stress
  σ_m = mean stress, load divided by area
  ϵ_B = strain at failure

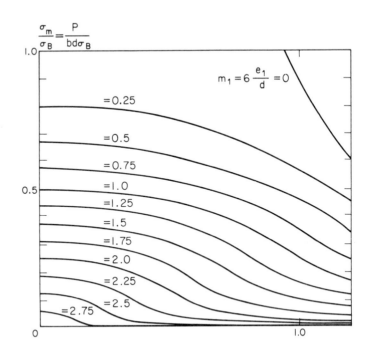

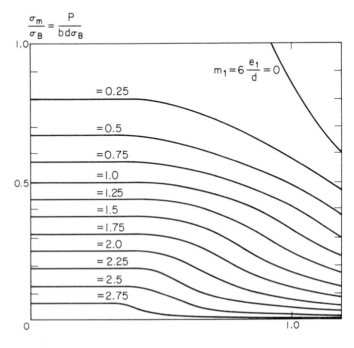

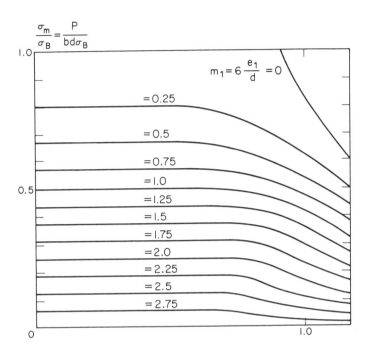

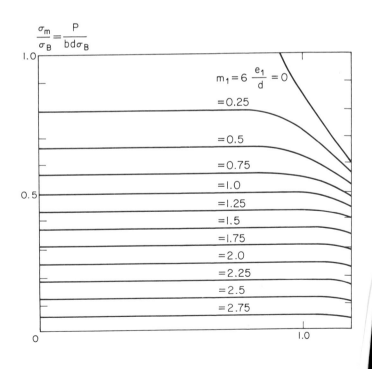

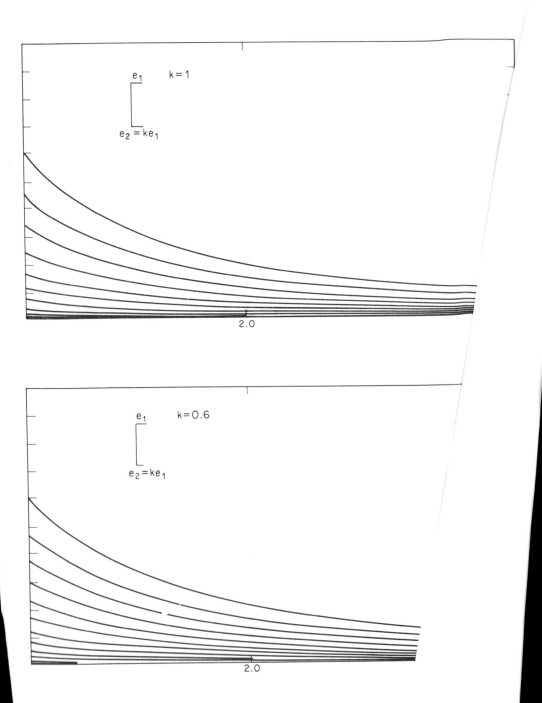

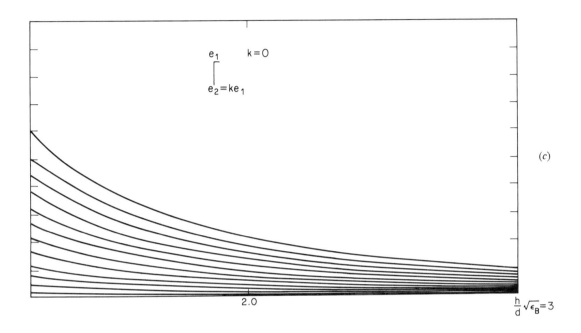

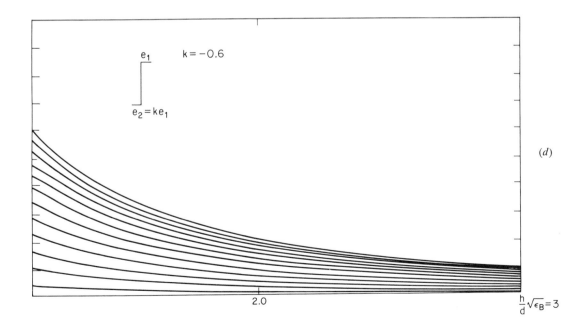

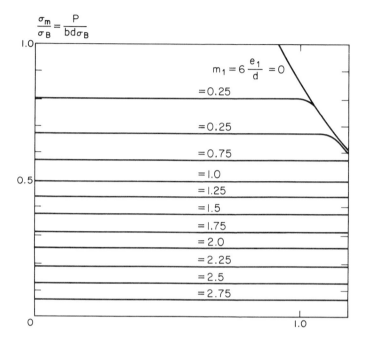

The stress-strain diagrams observed in tests have the general shape shown in Fig. E.13a. To make possible a comparison with the simple buckling formula (in Section E.10), a varying tangent modulus E_t was defined, as shown in Fig. E.13b. The simple buckling formula, Eq. (E.1), does not take into account eventual material failure, nor does it take into account eventual eccentricities of the axial load. These effects are acounted for in Figs. E.12, which are, however based upon the assumption that the stress-strain diagram is straight, as shown in Fig. E.13c. The strain ε_B can now be taken to be as shown in Fig. 13d, the larger value underestimating the stiffness of the wall except for at stresses close to the ultimate, the lower value overestimating the stiffness except for very low stresses where the approximation is best fitted. These two values from the test data would define two values for ε_B, one giving an upper limit for the buckling load and the other giving a lower limit except for close to the ultimate stress of the material, where the stress-strain diagram is not too well defined anyway. From the stress-strain diagrams taken from test series studied in Fig. D.15, an approximate value of ε_B can be found [E.7]: ε_B is from 0.002 to 0.0045. If a relatively high intermediate value is taken, say $\varepsilon_B = 0.004$, a h/d ratio of 63

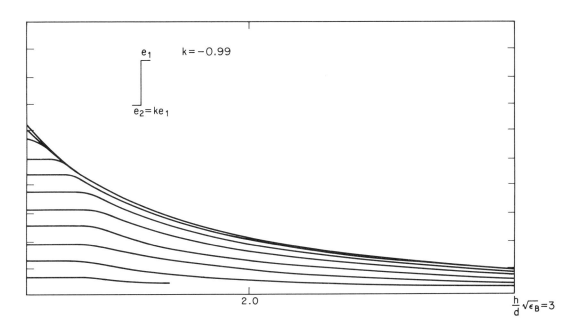

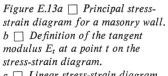

Figure E.13a □ Principal stress-
strain diagram for a masonry wall.
b □ Definition of the tangent
modulus E_t at a point t on the
stress-strain diagram.
c □ Linear stress-strain diagram
assumed in the theoretical
calculations.
d □ Application of Fig. E.13c to
Fig. E.13a.

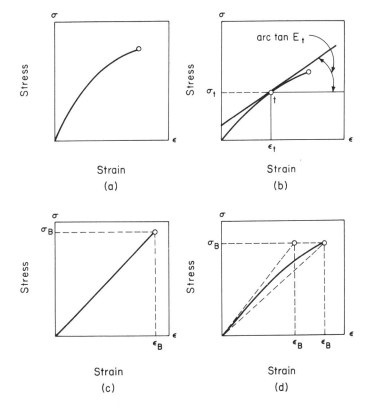

will correspond to an $h/d \sqrt{\varepsilon_B}$ value of 1.0 in Fig. E.12a, since Fig. E.12 was prepared for pin-ended columns, while the test specimens reported in Fig. D.15 were fixed at both ends.

If it is also assumed that an unintended end eccentricity of 4% of the wall thickness was present in the tests, the curve marked $m_1 = 0.25$ in Fig. E.12a would apply. This is probably an over-estimation of the eccentricity in some of the tests, but the overall shape of the theoretical curve as compared with the test data in Fig. D.15 generally fits the test results well (Fig. E.14). The theoretical curve for no eccentricity has also been shown in the figure. It should be remembered when making a comparison that the eccentricity is unknown and the stiffness in bending as represented by ε_B is also unknown from the tests and had to be estimated.

Our comparison has served a two-fold purpose: to compare theory and test in principle, and to demonstrate the use of Figs. E.12a to E.12e.

Figure E.14 □ *Comparison of theoretical and experimental data, only an example. See explanation in text.*

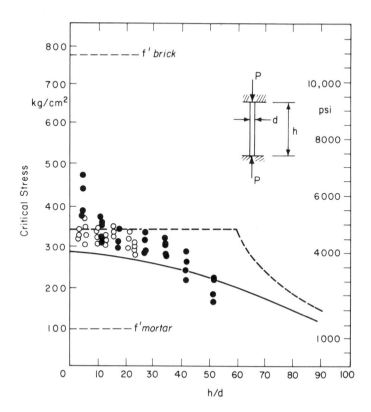

References for Chapter E

E.1 ☐ Angervo, K.: "Über die Knickung und Tragfähigkeit eines exzentrisch gedrückten Pfeilers ohne Zugfestigkeit." (On the Buckling and the Bearing Capacity of an Eccentrically Compressed Pillar with No Tensile Strength.) Staatliche Technische Forschungsanstalt, Finnland, Publication 26, Helsinki, 1954.

E.2 ☐ Chapman, J. C., and Slatford, J.: "The Elastic Buckling of Brittle Columns." Paper No. 6147, Proceedings of the Institution of Civil Engineers, London, Vol. 6, pp. 107–125, January, 1957.

E.3 ☐ Angervo, K. and Putkonen, A.: "Erweiterung der Theorie der Biegung eines Pfeilers ohne Zugfestigkeit." Staatliche Technische Forschungsanstalt, Finnland, Publication 34, Helsinki, 1961.

E.4 ☐ Sahlin, S.: "Structural Interaction of Walls and Floor Slabs." Inst. för Byggnadsstatik KTH, Meddelande Nr. 33, Stockholm, 1959; Handling No. 35, National Swedish Council for Building Research, Stockholm, 1959.

E.5 ☐ Nylander, H.: "Undersökning av Bärkraften hos Murade cementstensväggar." (Investigation of Load-Carrying Capacity of Cement Block Masonry Walls.) Betong, Häfte 3, Stockholm, 1944.

E.6 ☐ Sahlin, S.: "Diagrams of Critical Stress for Columns of Material without Tensile Strength." National Swedish Institute for Building Research, Report 16/65, Stockholm, 1965.

E.7 ☐ Structural Clay Products Research Foundation: "Compressive Strength Tests of Four-inch Brick Walls." Research Report No. 9, Geneva, Illinois, 1965.

F □ The interaction of walls and floor slabs: joint behavior

F.1 □ Introductory remarks

The load-carrying capacity of a wall is strongly dependent on the eccentricity of the load. If the wall is connected to a loaded slab, the eccentricity will depend on the rigidity of the slab and the joint as well as on the rigidity of the wall. The load eccentricity can be calculated theoretically under the condition that the structure is continuous. The continuity at the joints requires that the angle of rotation of the wall end plus the rotation in the joint itself equals the end rotation of the slab (see Fig. F.3).

$$\varphi_v + \theta = \varphi_h \tag{F.1}$$

A fundamental point in the theory is the behavior of the joints which will be treated in more detail below. The responses of the slabs and the walls are, of course, also relevant and are included in the discussion.

F.2 □ The deformation (end rotation) of the wall

In Chapter E it was shown that the angle of rotation of a wall having no tensile strength depends on the magnitude and the eccentricity of the axial load. (See Fig. E.8 and related text.) If the wall is uncracked and linearly elastic, the solution can be found in most textbooks on structural mechanics. See Timoshenko [F.1].

Generally speaking, it can be said that the end rotation φ of a wall is a function of P and e.

$$\varphi_v = f(P, e) \tag{F.2}$$

where P is the axial load and e is the end eccentricity of the load. It may also be observed that the moment acting at the wall end is

$$M_v = Pe \tag{F.3}$$

Notations in Chapter F

A = area of cross section

$E_h I_h$ = flexural rigidity of a horizontal structural part (floor slab or horizontal frame member)

$E_v I_v$ = flexural rigidity of a vertical structural part (wall or vertical frame member)

$G = k(qL/2)$ = dead load per story due to the weight of a wall

L = length of span of a horizontal structural part (floor slab or horizontal frame member) = distance between the center lines of vertical structural parts (walls or vertical frame members)

M = bending moment

M_{pl} = bending moment in the case of fully developed plastification

M_v = moment acting at the wall end

P = compressive force

P_c = ultimate load of a vertical structural part (wall) submitted to a central load

P_l = compressive force acting on a vertical member immediately below a floor slab

P_u = compressive force acting on a vertical member immediately above a floor slab

c = coefficient used in the calculation of the angle of rotation at an end of a wall, see Eq. (F.29)

d = thickness of a vertical structural part (wall or vertical frame member)

e = eccentricity of the compressive force acting on a vertical structural part (wall or vertical frame member)

e_l = eccentricity of the compressive force acting on a vertical member immediately below the floor slab

e_u = eccentricity of the compressive force acting on a vertical member immediately above the floor slab

$f(P, e)$ = function of P and e

h = total height of a vertical structural part (wall or vertical frame member) in a story

$k = G/[q(L/2)]$

$m = 6(e/d)$ = relative eccentricity of the compressive force acting on a vertical structural part

m_1 = relative eccentricity of the compressive force at the top end of a vertical structural part

m_2 = relative eccentricity of the compressive force at the bottom end of a vertical structural part

n = ordinal number of a floor slab reckoned from the topmost story of a building

q = total uniformly distributed load per unit length acting on a floor slab

θ = total angle of rotation of the joint between a horizontal structural part and a vertical structural part, i.e. the difference between the angle of rotation of an end of the horizontal part and the angle of rotation of the vertical part at the joint

κ = ratio of the ultimate load of a wall subjected to an eccentric load to the ultimate load of the same wall under the action of a central load

λh = height of a vertical structural part (wall or vertical frame member) from a floor slab to the nearest point of inflection (zero moment point) below this floor slab

σ = stress (compressive stresses are usually reckoned as positive)

σ_c = ultimate stress under the action of a central load

σ_{ult} = ultimate stress; σ_{edge} = edge stress, etc.

φ_h = angle of rotation of an end of a horizontal structural part (floor slab or horizontal frame member)

φ_v = angle of rotation of an end of a vertical structural part (wall or vertical frame member)

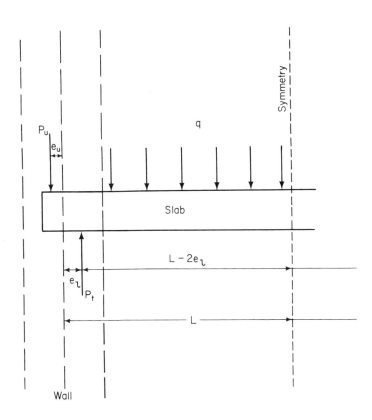

Figure F.1 ☐ Loads acting on a floor slab which forms part of a load-carrying structure of the type shown in Fig. F.3.

F.3 ☐ The deformation (end rotation) of the slab

A slab deflects when loaded, and its end rotates. This rotation is counteracted by the moment acting at the slab end. Consider a symmetrical slab loaded as indicated in Fig. F.1. The slab end is loaded from above by a load P_u in the wall. This load has the eccentricity e_u measured from the center line of the wall. Similarly, the total force in the wall below the slab is P_l and has the eccentricity e_l. Finally, the dead load and the live load are assumed to be uniformly distributed and their sum equal to q.

Under these assumptions the slab end will rotate:

$$\varphi_h = \frac{qL^3}{24E_hI_h} - M\frac{L}{2E_hI_h} = \frac{qL^3}{24E_hI_h} - (P_ue_u + P_le_l)\frac{L}{2E_hI_h}$$

(F.4)

if $\quad e_l \ll L$

In the uppermost story, Eq. (F.4) is simplified to

$$\varphi_h = \frac{qL^3}{24E_hI_h} - \frac{P_le_lL}{2E_hI_h} \tag{F.5}$$

Since the reaction at the support of the uppermost slab is

$$P_l = \frac{qL}{2} \tag{F.6}$$

we obtain

$$\varphi_h = \frac{P_lL^2}{12E_hI_h} - \frac{P_le_lL}{2E_hI_h} = \frac{P_lL^2}{12E_hI_h}\left(1 - 6\frac{e_l}{L}\right) \tag{F.7}$$

and thus the angle of rotation of the slab end is a function of P and e:

$$\varphi_h = f(P_l, e) \tag{F.8}$$

In an intermediate story, the magnitude of the normal force in the wall is, for the nth floor slab reckoned from the top of the building,

$$P_u = (n - 1)\frac{qL}{2} + (n - 1)G \tag{F.9}$$

where G is the dead load of the wall per story. Below the slab the reaction from the slab has to be added. Thus

$$P_l = P_u + \frac{qL}{2} = n\frac{qL}{2} + (n - 1)G \tag{F.10}$$

If we assume that the wall weighs a fraction k of the slab reaction, then

$$G = k\frac{qL}{2} \tag{F.11}$$

By inserting Eqs. (F.6), (F.10), and (F.11) in Eq. (F.4), and after some algebraic manipulations, the following equation is obtained:

$$\varphi_h = P_l\frac{L^2[1 - 6(n + nk - k - 1)e_u/L - 6(n + nk - k)e_l/L]}{12E_hI_h(n + nk - k)} \tag{F.12}$$

which again shows that the angle of rotation is a function of the load and the eccentricity for a given system.

$$\varphi_h = f(P_l, e_l, e_u) \tag{F.13}$$

If $e_l \approx e_u \approx e$, which often is the case, and with $P_l = P$, which is the most important force at the slab end (since the final failure occurs below the slab end), the equation can be rewritten

$$\varphi_h = f(P, e) \tag{F.14}$$

Similar derivations can be made, at least approximately, for almost any type of building system with load-bearing walls by common methods of structural mechanics.

F.4 □ Deformation and strength of joints

It is known from tests that a joint between a concrete slab and a masonry wall is inelastic above a certain moment (Sahlin [F.2]). Investigations by Emperger [F.3] on concrete beams, and by Kazinczy [F.4] on steel beams which were built in at the supports in brick masonry walls, corroborate the assumption that the behavior of most types of joints between a horizontal member and a masonry wall is to some degree inelastic. After the joint has started to rotate, the moment at the slab ends stays fairly constant until the rotation reaches a limit, which is accompanied by extensive vertical cracking above the slab end and by compression cracks (splitting) in the wall material under the slab end below the support area (Fig. F.2). Since the moment stays nearly constant, the load eccentricities (e_u and e_l in Fig. F.1) decrease with increasing load.

The failure of the joint is principally separated from the failure of the wall, but the ultimate load can also be determined by the failure of the wall a few bricks or blocks below the slab where the stress disturbances from the joint with the slab have disappeared. Thus, the joint can fail as described above and in Fig. F.2, while the wall is still intact and low stresses are fairly linearly distributed in the wall a few blocks or bricks below the slab end, or the joint can stay unbroken while the wall fails in crushing a few blocks or bricks below the slab end where the confining stresses from the slab have diminished.

An approximation of the behavior of the joint has been suggested [F.2] to be as follows:

$$\theta = 0 \qquad \text{for } 0 \leq M < M_{pl}$$
$$0 \leq \theta \leq \theta_{ult} \qquad \text{for } M = M_{pl} \tag{F.15}$$

where θ is the rotation which takes place in the joint itself, M is the moment at the slab end acting on the joint, M_{pl} is the moment at which the joint starts to act inelastically (plastic moment), and θ_{ult} is the rotation at which the joint finally fails. The assumptions are presented graphically in Fig. F.3.

Tests [F.2] on statically indeterminate systems (with frame action, Fig. F.4) showed a very pronounced yielding or inelastic behavior at a certain moment for some specimens (Fig. F.5).

Figure F.2 □ Photograph taken after failure of a joint between a concrete slab stub and a piece of a masonry wall without tensile strength, consisting of light weight cellular concrete blocks joined by plastic discs and no mortar. Clay brick and concrete block masonry with mortar show similar failure mechanisms (Swedish National Institute for Materials Testing, Intyg U64-3778).

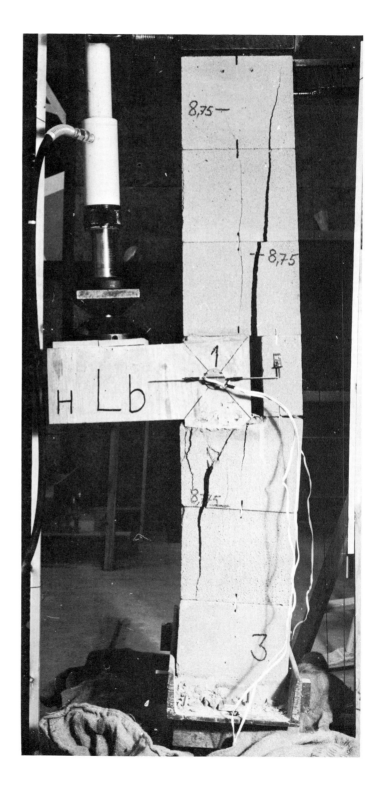

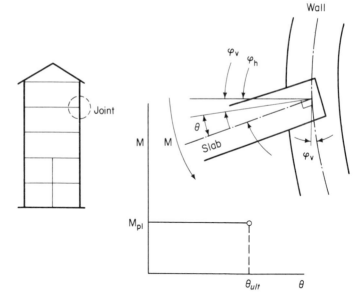

Figure F.3 ☐ (Right) Type of joint and idealized relationship between applied moment M and angle of rotation θ in a joint. θ_B equals ultimate rotation. M_pl equals "yielding" moment.

Figure F.4 ☐ (Below) Full scale test specimen simulating external walls in an intermediate story in a multistory building. All distances are in millimeters (divide by 25.4 to obtain inches) [F.2].

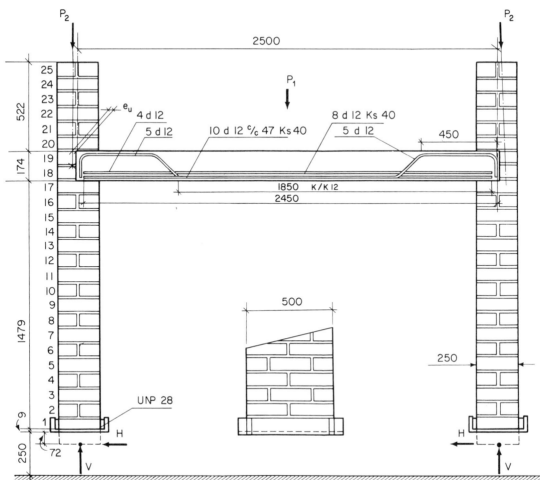

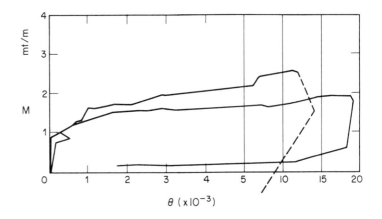

Figure F.5 ☐ Relationship between the moment acting on a slab end (and on the joint) and the angle of rotation in the joint. Brick walls built with lime-cement-sand mortar, ratio 2:1:15 [F.2].

Figure F.6 ☐ Type of full scale test specimen simulating a joint in an intermediate story in a multistory building (Swedish National Institute for Materials Testing, Intyg U64-3778).

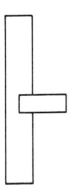

Note a rotation in the joint of approximately 1.5% before failure.

Tests conducted under the auspices of a committee established by the National Swedish Board for Building and Planning to devise codes for masonry walls [F.11], [F.5] showed, for some statically determinate specimens (Fig. F.6), a less pronounced "yield plateau," exemplified in Fig. F.7. The rotation was about 2% ($\theta \approx \tan \theta = 0.02$) at failure and about 0.8% at maximum moment. This latter moment-rotation diagram (Fig. F.7) does not exactly fit the idealized curve in Fig. F.3. In a practical case, however, it is possible to make several calculations with different assumptions for M_{pl} and θ_{ult} to fully account for the behavior of the joint type in question (whose behavior has to be tested or estimated from earlier tests); see Fig. F.8, which demonstrates the flexibility of the assumption according to Eq. (F. 15).

A more detailed M–θ curve could be assumed, at the expense of simplicity in calculations. With the relatively few test data obtained so far to support such an assumption, the eventual gain in precision is doubtful.

The "yield" moment M_{pl} and the rotation θ_{ult} at failure can both, theoretically, have any value in any combination. The values are within definable limits however, for common types of joints: M_{pl} ranges from 0.5 to 5mt/m (1 ft kips/ft to 10 ft kips/ft); and θ_{ult}, from 0 to 3%. The rotation θ_{ult} is influenced by the axial load in the wall, which is understandable in light of the fact that a centrally loaded wall loaded very close to failure will tolerate hardly any disturbance at all due to joint rotations. On the other hand, if the slab is resting on the top of a wall, as

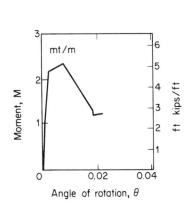

Figure F.7 □ *Relationship between moment and angle of rotation of joint, observed on one test specimen of the type shown in Fig. F.6 (Swedish National Institute for Materials Testing, Intyg U64-3778).*

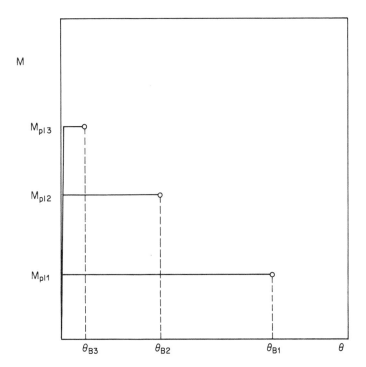

Figure F.8 □ *Example of moment rotation relationships obtainable with the approximate relationship of Eq. (F.26).*

Figure F.9 □ *Angle of rotation of a joint at failure θ_{ult} as a function of the relative axial load in the wall $\kappa = P/P_c$, a theoretical approximation.*

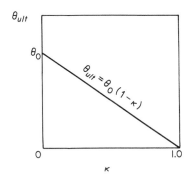

the topmost slab in a building does, and this slab does not carry any load, the permissable rotation is almost unlimited but cannot be utilized for practical reasons.

If the load-carrying capacity for a concentrically loaded wall is called P_c, the relative failure load for an eccentrically loaded wall can be denoted by $\kappa = P/P_c$ ranging from 0 to 1. As κ increases, θ_{ult} decreases. A simple expression for this relationship is

$$\theta_{ult} = \theta_0(1 - \kappa) \tag{F.16}$$

where θ_0 is the intercept on the θ axis in the diagram, Fig. F.9.

Tests [F.2] and [F.5] show such a relationship between θ_{ult} and κ, although the scatter is appreciable, and the value of θ_0 varies with the type of the joint and the materials in the wall and in the slab. For 10-in. brick masonry walls with 5-in. slab support, θ_0 is about 0.03; and for 10-in. walls and $2\frac{1}{4}$-in. support length, θ_0 is about 0.1: for 10-in. light weight cellular concrete walls with 6-in. slab support, θ_0 is in the neighborhood of 0.04.

Table F.1 ☐ Left half; right half on facing page.

	Source	Type of specimen	Wall			Slab		
			Material	Thickness (cm)	Approximate strength (kg/cm²)	Material	Thickness (cm)	Approximate strength (kg/cm²)
1	[F.2]	frame	brick masonry	25	70	concrete	22.0 or 17.4	350
2		"	"	25	45	"	17.4	350
3		"	"	25	70	"	17.4	350
4	[F.11]	joint	"	25	75	"	16	
5		"	"	12	75	"	16	
6		"	"	12	75	"	16	
7		"	lt. wt. cellular concrete	25	22	"	16	
8		"	lt. wt. aggregate concrete	25	30	"	16	
9		"	hollow concrete block	20	50	"	16	
10		"	brick masonry	12	75	"	16	

For some other joint types [F.5], θ_0 is 0.03 to 0.1. See Table F.1.

Since $M = P_u e_u + P_l e_l$, as Fig. F.1 shows, the moment can be expressed as a function of P and e with the aid of Eqs. (F.5), (F.6), and (F.11):

$$M = P_l \left(\frac{n - 1 + kn - k}{n + kn - k} e_u + e_l \right) \tag{F.17}$$

or if $e_u \approx e_l \approx e$ and $P = P_l$,

$$M = Pe \frac{2(n + nk - k) - 1}{n + nk - k} = PeF(n, k) \tag{F.18}$$

which may be substituted in Eq. (F.15) to give the following approximate relationship between the angle of rotation and the

Table F.1 □ Inelastic moments M_{pl} and joint rotations θ observed in tests. The data are condensed from the sources listed in the table; the range of actual test values can be found in the detailed descriptions in the original sources.

| | Story | | M_{pl} | M_{pl}/Pc | θ_0 | Support | Number of |
	Top	Internal	(ton-m/m)	(cm)		length (cm)	observations
1	×		2.5	1.4	approx. 0.03	12	4
2		×	2.1	2.1	0.03	12	8
3		×	4.7	2.7	0.03	12	8
4		×	1.6		0.1	6	1
5		×	1.0		0.03	12	2
6	×		0.9		0.06	12	1
7		×	1.2–2.3	0.8–2.2	0.04	15	4
8		×	3.0	2.2	0.1	15	1
9	×		1.0		0.03	15	1
10	×		0.9		0.05	12	1

eccentricity:

$$\theta = 0 \qquad \text{for } 0 \le PeF(n, k) \le M_{pl}$$
$$0 < \theta \le \theta_{\text{ult}} \qquad \text{for } PeF(n, k) = M_{pl} \tag{F.19}$$

Thereby it has been shown that θ is a function of P and e,

$$\theta = f(P, e) \tag{F.20}$$

for a given slab n and a given weight of the wall k. The eccentricity e can, at the same time, be a function of P since it decreases with increasing load in the inelastic stage.

F.5 □ Interaction of walls, joints, and floor slabs

It has been shown in the foregoing sections that the angles of rotation of the wall end φ_v, the slab end φ_h, and the joint θ can

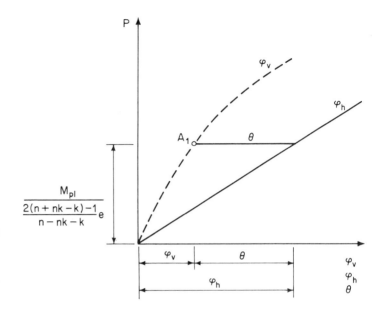

Figure F.10 □ Principal description of a graphical method to obtain the relationship between the applied load P and the end rotation of a wall φ_v, as well as the eccentricity of the load e. See discussion in text.

be expressed as functions of P and e. The φ_h–P relationship for different values of e can be plotted in the same diagram as the relationship for φ_v–P (which was shown in Fig. E.8). The lack of fit between the φ_h and φ_v curves has to be compensated by θ as required by Eq. (F.1). The principle is shown in Fig. F.10. The curve φ_v follows Eq. (F.2) and is shown for only one eccentricity, e. A family of curves for several values of e is shown in Fig. E.8. The curve φ_h representing the angle of rotation of the slab end has been plotted according to Eq. (F.14) for the eccentricity which was chosen for the φ_v curve. The lack of fit which can be calculated from Eq. (F.1) between these two curves is the rotation θ in the joint itself. The rotation θ can be calculated from the second of Eqs. (F.19) with the moment according to Eq. (F.18), and the Eqs. (F.1), (F.2) and (F.14). Again the same eccentricity e must be used throughout.

It has been assumed that the joint is in the inelastic stage. If not, the rotation θ diminishes to zero.

The point A_1 in the diagram represents an eccentricity e_1 and a load P_1 for which the conditions of Eqs. (F.1), (F.2), (F.14), and (F.19) are all fulfilled for the same e. The process is then repeated for another eccentricity, e_2, which gives another load P_2, and a new point A_2, in the coordinate system P–φ_v, and so on.

The points A_1, A_2, etc., as determined above, can be joined by a curve which shows the relationship between the load P and the angle of rotation of the wall end φ_v, and along the curve the variation of the eccentricity is known. The parameters of interest, P and e, are now known for the whole loading process.

The maximum load is reached when the ultimate edge stress in the wall is reached, when θ_{ult} is reached, or when the slab fails, whichever occurs first. Suppose that the wall strength governs the failure. We then have to calculate the maximum edge stress in the wall. This stress can be obtained from the following equations if the stress distribution in the wall follows Fig. F.11. When the load acts inside the kern, the edge stress is

$$\sigma_{edge} = \frac{P}{A}\left(1 + 6\frac{e}{d}\right) \qquad \text{for } 0 \leq \frac{e}{d} \leq \frac{1}{6} \tag{F.21}$$

and when it acts outside the kern, the edge stress is

$$\sigma_{edge} = \frac{4P}{3A(1 - 2(e/d))} \qquad \text{for } \frac{1}{6} \leq \frac{e}{d} \leq \frac{1}{2} \tag{F.22}$$

The ultimate load is reached when σ_{edge} reaches the ultimate stress. It is possible to draw curves of equal stress in the diagram showing the load-rotation relationship of the wall, as has been done in Fig. E.8. When that particular curve of ultimate stress which applies to the actual material and dimensions is reached

Figure F.11 □ *Stress distribution when the axial load is inside the kern (left) and outside the kern (right) of a wall of a material having no tensile strength.*

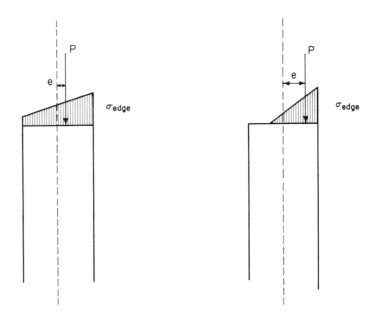

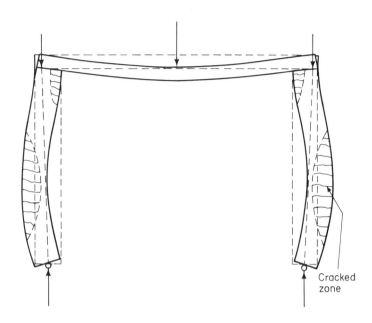

Figure F.12 ☐ Exaggerated deflections at the moment of buckling of a frame consisting of a single span concrete slab and two walls having no tensile strength.

Cracked zone

along the line A_1, A_2, etc., just obtained, the wall fails—if the joint has not already failed because θ_{ult} was reached. As a matter of fact, trial calculations show that M_{pl} is not very often reached for common structures with two-way slabs.

There is also the possibility that the ultimate stress never is reached but the wall becomes unstable. This is possible only for very slender walls (without lateral load). The problem then becomes quite intricate and has been solved only for a few cases. In Fig. F.12 the final stage of the loading process for a simple frame is shown in principle, with the eccentricity reversed at the wall end and an ever-increasing eccentricity at about mid height of the wall, where the wall cracks more and more and finally is reduced in section so much that the load passes a maximum equal to the ultimate load. For further details, see Hellers and Sahlin [F.6] and [F.2], page 28. The safety factor for buckling is usually high for one-story-high walls, and the restraints on a wall offered by adjacent members are probably even more beneficial than for regular columns with high tensile strength [F.6]. Although it is fully possible to calculate the load-carrying capacity of a wall in the way described above, shortcuts and approximations can be employed in cases where the behavior of the structure during the whole loading process is of no interest.

F.6 □ Graphic solution using a dimensionless diagram

The curves showing the relationship between the load P and the end rotation under constant edge stress can be redrawn in another system of coordinates, as shown in Fig. F.13. On the curves, the relative eccentricities e/d are marked with dashed lines. The curves all lie in a comparably narrow region. This fact will be used later on for developing an approximate method of design.

The relation between κ and $(\varphi_h - \theta)$ can be plotted in the same coordinate system. This relation is dependent on the relative eccentricity e/d, among other factors, as expressed by Eq. (F.23).

$$\kappa = \left(\frac{12E_h I_h \lambda hd}{E_v I_v L^2} \frac{E_v I_v}{P_c \lambda hd} \varphi_v + \frac{12E_h I_h}{P_c L^2} \theta\right)$$

$$\times \left(\frac{n + nk - k}{1 - 6(n + nk - k - 1)\frac{e_u}{L} - 6(n + nk - k)\frac{e_l}{L}}\right)$$

(F.23)

If θ, e_u, and e_l are constant, Eq. (F.23) represents a straight line in the coordinate system shown in Fig. F.13.

The value of θ depends on the stage of loading under consideration. Three cases can be met.

Case 1:

$$\theta = 0 \quad \text{if} \quad M = P_l\left(\frac{n + nk - k - 1}{n + nk - k} e_u + e_l\right) \leq M_{pl}$$

(F.24)

Different values of e are inserted in Eq. (F.23) until a point on the curves in Fig. F.13 is found for which e is the same both in Eq. (F.23) and Fig. F.13. If Eq. (F.24) is not fulfilled. Case 2 should be examined.

Case 2:

$$M = M_{pl} \quad \text{and} \quad \theta \leq \theta_{\text{ult}}$$

gives

$$\kappa = \frac{M_{pl}}{P_c\left(\frac{n + nk - k - 1}{n + nk - k} e_u + e_l\right)}$$

(F.25)

Different values of e are inserted in Eq. (F.25), and the procedure for Case 1 is repeated.

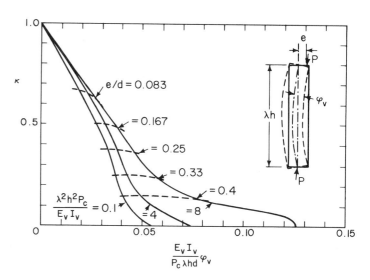

Figure F.13 ☐ Relation between the angle of rotation of the wall φ_v and the relative value of the ultimate load κ (κ = the ratio of the ultimate load in the case of eccentric loading to the ultimate load in the case of concentric loading.) This relation is represented by full-line curves. The relative eccentricity of the load e/d is represented by dash-line curves. The eccentricity of the load is e at one end of the wall and 0 at the other end.

If the calculated value of κ corresponds to $\theta > \theta_{\text{ult}}$, Case 3 must be checked. The value of θ can be computed from Eq. (F.23) when e is known, since φ_v can be taken from Fig. F.13.

Case 3: The rotation φ_v as a function of the load is assumed to be represented by a straight line which passes through the origin of the coordinate system and through a point on the κ–φ_v curves corresponding to the proper values of e and $P\lambda^2h^2/E_vI_v$. The angle φ_h can be determined by Eq. (F.23) on the assumption that $\theta = 0$; that is, $\varphi_v = \varphi_h$. The difference between these two angles of rotation (at the same value of e) is equal to the angle which is to be found (θ) if Eq. (F.25) is satisfied at the same time. The value of θ_{ult} is found from Eq. (F.16).

This again gives the necessary constituents for a calculation of load-carrying capacity. By using a trial and error method, we can calculate the value of e/d in each individual case, and then the relative value κ for the ultimate load P_{ult} is seen directly.

F.7 ☐ Approximate explicit solution

The curves in Fig. F.13 which represent the relationship between κ and φ_v can be approximated by a single straight line expressed by the equation

$$\kappa = 1 - \frac{\dfrac{E_vI_v}{P_c\lambda hd}}{k_4}\varphi_v \qquad (F.26)$$

where the value of k_4 is about 0.05 if the approximation is chosen on the safe side of the family of curves.

By elimination of φ_v from Eqs. (F.23) and (F.26), a relationship between κ and e may be obtained. If the maximum stress at failure is the same for a concentrically and an eccentrically loaded wall, then the product $A \cdot \sigma_{\text{edge}}$ [see Eqs. (F.21) and (F.22)] could be thought of as the load P_c for which a concentrically loaded wall fails. After some simple derivations, Eqs. (F.21) and (F.22) give a second relationship between κ and e.

$$\frac{P}{P_c} = \kappa = \frac{3}{4}\left(1 - 2\frac{e}{d}\right) \qquad \text{for } \frac{1}{6} \le \frac{e}{d} \le \frac{1}{2}$$

$$\frac{P}{P_c} = \kappa = \frac{1}{1 + 6\dfrac{e}{d}} \qquad \text{for } 0 \le \frac{e}{d} \le \frac{1}{6} \tag{F.27}$$

From the two mentioned relationships between κ and e, κ can be calculated and expressed explicitly [F.2]. Since there are two distinct cases of the location of the axial load [of Eq. (F.27)] and three distinct modes of failure regarding the joints,

1. $M = M_{pl}, \quad \theta = 0 \quad \sigma = \sigma_{\text{ult}}$
2. $M = M_{pl}, \quad \theta < \theta_{\text{ult}} \quad \sigma = \sigma_{\text{ult}}$ (F.28)
3. $M = M_{pl}, \quad \theta = \theta_{\text{ult}} \quad \sigma < \sigma_{\text{ult}}$

six different equations are obtained when all the combinations are considered. These equations are given in Table F.2. In this table one more case of joint action has been added: the case when the yield point of the joint is less pronounced and the yielding takes place progressively. As a limit for the load in such a case, the maximum permissible rotation alone has been taken as a condition for failure:

$$\theta_{\text{ult}} = \theta_0(1 - \kappa) \tag{F.16}$$

The theoretical values were compared with test data [F.2] on eight frames of the general type shown in Fig. F.4, and the agreement was good.

The κ values have been computed by the method mentioned above for several values of the parameters involved: material behavior, dimensions, etc. These calculations are reported in [F.7]. The results are of the general character shown in Fig. F.14. The diagram in Fig. F.14, however, applies only to the parameters shown in the figure. With a set of such diagrams calculated from the equations presented above, intermediate spans, wall stiffnesses, and so forth, can be determined by interpolation.

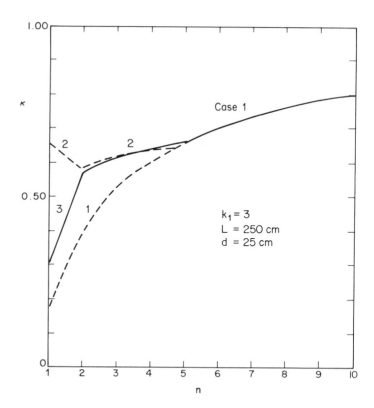

Figure F.14 □ *Ultimate loads of brick masonry walls in different stories expressed as a fraction of the ultimate load of a wall subjected to a concentric load,* $k_1 = 12E_hI_h\lambda \, hd/E_vI_vL^2$.

Since diagrams of this sort apply to a certain type of structure [shown in Figs. F.3 and F.4] with single span slabs only, equivalent spans of the slabs must be calculated if the slab is supported on, for example, all four sides.

Constants for the formulas in Table F.2:

$$P_c = \sigma_B bd$$

$$k_1 = \frac{12E_hI_h\lambda hd}{E_vI_vL^2}$$

$$k_2 = \frac{E_vI_v}{P_c\lambda hd}$$

$$k_3 = \frac{12E_hI_h}{P_cL^2}$$

$$k_4 = 0.05$$

$$k_5 = 0.375 - \frac{1 + 0.05k_1(1.2n - 0.2)}{16\dfrac{d}{L}(1.2n - 0.7)}$$

$$k_6 = \frac{0.05k_1(1.2n - 0.2)}{8\dfrac{d}{L}(1.2n - 0.7)}$$

108

$$k_7 = 0.05k_1(1.2n - 0.2) + 2\frac{d}{L}(1.2n - 0.7)$$

$$k_8 = \frac{M_{pl}(1.2n - 0.2)}{2P_c d(1.2n - 0.7)} \quad \left(\geq \frac{3}{32} \text{ indicates Case 1}\right)$$

$$k_9 = \frac{k_2\theta_0 + k_8\left[\dfrac{12\dfrac{d}{L}(1.2n - 0.7) + 2}{k_1(1.2n - 0.2)} + 2(0.05 + k_2\theta_0)\right]}{2\left[\dfrac{1}{k_1(1.2n - 0.2)} + k_2\theta_0 - \dfrac{0.05}{3}\right]}$$

$$k_{10} = \frac{2k_8\left[k_2\theta_0 + \dfrac{12\dfrac{d}{L}(1.2n - 0.7)k_8}{k_1(1.2n - 0.2)}\right]}{\dfrac{1}{k_1(1.2n - 0.2)} + k_2\theta_0 - \dfrac{0.05}{3}}$$

$$k_{11} = 0.375 - \frac{1 + 1.3k_2\theta_0 k_1(1.2n - 0.2)}{16\dfrac{d}{L}(1.2n - 0.7) + 8 \cdot 0.05k_1(1.2n - 0.2)}$$

$$k_{12} = \frac{k_2\theta_0 k_1(1.2n - 0.2)}{8\dfrac{d}{L}(1.2n - 0.7) + 4 \cdot 0.05k_1(1.2n - 0.2)}$$

$$k_{5b} = 0.375 - \frac{1 + (0.05 + k_2\theta_0)k_1(1.2n - 0.2)}{16\dfrac{d}{L}(1.2n - 0.7)}$$

$$k_{6b} = \frac{(0.05 + k_2\theta_0)k_1(1.2n - 0.2)}{8\dfrac{d}{L}(1.2n - 0.7)}$$

$$k_{7bm} = (0.05 + k_2\theta_0)k_1(1.2n - 0.2) + 2\frac{d}{L}(1.2n - 0.7)$$

The variation in load-carrying capacity κ by a variation of k_1, which is a major parameter in the problem expressing the stiffness ratio of slab to wall, has been studied by Awad [F.10] with the aid of a programmed version of Table F.2 run on an IBM 7094 plus a Calcomp plotter. The principal results are exemplified in Fig. F.15, showing κ for a variation of k_1 from 1 to 5.

Figure F.15 shows that an increase in slab stiffness substantially improves the load-carrying capacity of the wall, except in cases where the joint is "yielding" and a change in slab stiffness does not decrease the deflection enough to prevent the "yielding," or increase the deflection enough to cause failure because of excessive deformation in the joint.

If it is doubtful whether the wall will crack or not, the constants k_{11} and k_{12} should be used instead of k_9 and k_{10} in order to check the danger of failure in the joint. [F.2]

Table F.2 ☐ Load-carrying capacity of masonry walls [F.7]. The constant k is assumed to have the value 0.2. See above for k_1 through k_{7bm}.

Case	$\kappa =$	(equation number)
1 $\theta = 0$ $M < M_{pl}$	$k_5 + \sqrt{k_5^2 + k_6}$	(32)
	$\dfrac{k_7}{1 + k_7}$	(33)
2 $0 < \theta < \theta_{ult}$ $M = M_{pl}$	$0.375 + \sqrt{0.375^2 - 1.5k_8}$	(36)
	$1 - 6k_8$	(37)
3 $0 < \theta = \theta_{ult}$ $M = M_{pl}$	$k_9 + \sqrt{k_9^2 - k_{10}}$	(40)
	$\dfrac{k_1(1.2n - 0.2)(k_2\theta_0 + 0.3k_8) + 12(d/L)(1.2n - 0.7)k_8}{1 + k_1(1.2n - 0.2)k_2\theta_0}$	(41)
3b $0 < \theta = \theta_{ult}$ $M = M_{pl}$	$k_{5b} + \sqrt{k_{5b}^2 + k_{6b}}$	(42)
	$\dfrac{k_{7bm}}{1 + k_{7bm}}$	(43)

	Validity interval	Check	(equation number)
	$0 \leq \kappa \leq 0.5$	If $$\frac{\kappa(3 - 4\kappa)(1.2n - 0.7)d}{3(1.2n - 0.2)} > \frac{M_{pl}}{P_c}$$ use Eq. (36) or (37).	(34)
	$0.5 \leq \kappa \leq 1$	If $$\frac{(1 - \kappa)(1.2n - 0.7)d}{3(1.2n - 0.2)} > \frac{M_{pl}}{P_c}$$ use Eq. (36) or (37).	(35)
	$0 \leq \kappa \leq 0.5$	If $$\frac{\kappa - 2(d/L)(1.2n - 0.7)(3 - 4\kappa)\kappa}{k_1(1.2n - 0.2)} - 0.05(1 - \kappa) > k_2\theta_0(1 - \kappa)$$ use Eq. (40) or (41).	(38)
	$0.5 \leq \kappa \leq 1$	If $$\frac{\kappa - 2(d/L)(1.2n - 0.7)(1 - \kappa)}{k_1(1.2n - 0.2)} - 0.05(1 - \kappa) > k_2\theta_0(1 - \kappa)$$ use Eq. (40) or (41).	(39)
$\dfrac{1}{6} \leq \dfrac{e}{d} = \dfrac{M_{pl}(1.2n - 0.2)}{2\kappa d P_c(1.2n - 0.7)} \leq \dfrac{1}{2}$			
$0 \leq \dfrac{e}{d} = \dfrac{M_{pl}(1.2n - 0.2)}{2\kappa d P_c(1.2n - 0.7)} \leq \dfrac{1}{6}$			
	$0 \leq \kappa \leq 0.5$		
	$0.5 \leq \kappa \leq 1$		

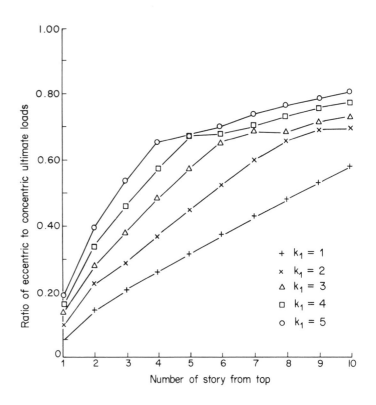

Figure F.15 ☐ Relative load-carrying capacity as a function of the ordinal number of the story from the top for varying values of k_1 which is the stiffness ratio of slab to wall:

$$k_1 = \frac{12 E_h I_h \lambda h d}{E_v I_v L^2}$$

The following assumptions were made:
The strength of masonry equals 50% of the brick strength, the wall modulus of elasticity equal 700 times the masonry strength, $\lambda = 0.5$, $k = 0.2$, $\theta_0 = 0.03$, $E_v = 700,000$ psi $(50,000\ kg/cm^2)$, $L = 240$ in. (6.1 m), $d = 10$ in. (25 cm), $\sigma_c = 1,000$ psi $(70\ kg/cm^2)$, $h = 120$ in. (3 m), $M_{pl} = 10,000$ lb in./in. (4536 kg cm/cm), $E_h = 3,000,000$ psi $(210,000\ kg/cm^2)$, $t = 6$ in. (15 cm) [F.10].

F.8 ☐ Alternative approximate methods of calculation

The problem of finding the load eccentricity in a wall can, of course, be attacked in many different ways just as the problem of frames of fully elastic material with high tensile strength. In our case, where the material in the vertical members has little or no tensile strength, the wall stiffness varies with the load and its eccentricity, and in addition the center line of the remaining intact wall becomes crooked. This is a complication.

The eccentricity effect can be considered by a reduction coefficient, as proposed by Nylander [F.8]. See Figs. E.9. and E.10. By assuming that the location of the inflection point of an external wall is fixed, the wall can be divided into statically similar pieces of the type shown in Fig. E.9. The end rotation of such a piece can be calculated from the modified equation for a member loaded by a moment at one end.

$$\varphi_v = \frac{P e \lambda h}{c(3 E_v I_v)} \tag{F.29}$$

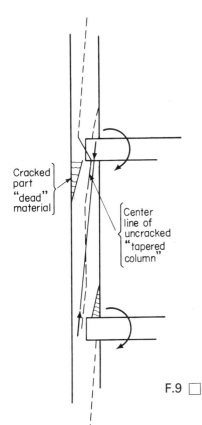

Figure F.16 □ *A wall having no tensile strength becomes "tapered" when loaded outside the kern due to cracking "dead material." In addition, the centroid axis of the remaining "active" part of the wall is bent.*

The factor *c*, which depends upon *e*, can then be read from the diagram in Fig. E.10. The computation is best performed by a trial and error method. An eccentricity e_n is assumed. The coefficient *c* is then found from Fig. E.10. The moment distribution in the load-carrying frame can then be calculated by any frame calculation method, since all stiffnesses are known. From the calculated moment *M* and the axial load *P*, the eccentricity $e_{n+1} = M/P$ is calculated; this gives a new value of *c*, and so on. The procedure has to be repeated until the difference $e_{n+1} - e_n$ is acceptable.

When *e* and *P* are known, the edge stresses can be computed and consequently P_{ult} is found. If the calculated moment in a joint exceeds M_{pl}, the eccentricity is overestimated, since *e* cannot exceed the value

$$e_{pl} = \frac{M_{pl}}{P} \tag{F.30}$$

This value of e_{pl} can be used for calculation of the wall load-bearing capacity provided

$$\varphi_h - \varphi_v = \theta_{ult} \le \theta_0 (1 - \kappa) \tag{F.31}$$

If the condition is not fulfilled, the joint fails because the rotation capacity is exceeded. The load must be chosen so that the condition is fulfilled.

Cases when the slab fails are not treated here.

F.9 □ Relaxation methods

With some modifications and allowances for the special behavior of masonry, relaxation methods such as the Cross method can be used to compute the moment distribution in a building structure consisting, for example, of walls plus slabs spanning between and connected to the walls. Where the axial load acts outside the core, the cracking will decrease the effective area and create what amounts to a "tapered column" (see Fig. F.16). Furthermore, the center line of the remaining uncracked part of the wall or column is polygon-formed, as indicated in Fig. F.16. Since the "tapering" cannot be known before the load eccentricities have been calculated, a trial and error process has to be added to the ordinary relaxation method.

Putkonen [F.9] has provided tables for the carry-over factors and stiffness factors (for different load eccentricities) for "tapered" columns of the type indicated in Fig. F.16. The process is, then, that by choosing an initial value of *e* at all ends of all walls the stiffness factors and carry-over factors are known and the relaxation of the fixed end moments can take place in a common manner. From the calculated moment distribution, the load eccentricities in all walls can be computed and new stiffness factors and carry-over factors can be determined. With this new set of factors the whole process is repeated until the differences between two consecutive sets of eccentricities are sufficiently small. The second-order effects are not taken into account.

F.10 ☐ Summary

The cracking of a masonry wall with low or no tensile strength, together with the inelastic action of the joints between the wall and the floor slab, causes calculation complications not encountered in ordinary fully elastic frames. The additional effects can be considered by a graphical method in which the load–deformation relationships of the wall, of the joint, and of the slab are repeatedly plotted simultaneously for the same load eccentricity. The result is a load-eccentricity-angle of rotation diagram. In the same diagram edge stresses for the wall are also plotted, enabling us to read edge stresses for increasing load, and since the ultimate stress is known for the actual material, the ultimate load is obtained. It is also possible to approximate the stress-deflection curves (Fig. F.13) and to obtain equations for direct calculation of the ultimate load in closed form from a set of six equations, varied for different failure mechanisms (Table F.2). A trial and error method can also be used in combination with a stiffness reduction factor, according to Figs. E.9 and E.10.

Alternatively, the moment distribution in a building frame consisting of slabs and walls can be calculated with a relaxation method, provided that the stiffness factors and carry-over factors are found by trial and error, because they vary as a function of the moments to be calculated.

Before one of the more time-consuming methods is started, it is advisable to make the following simple calculations and estimations: None of the known tests of the structures of the type shown in Fig. F.4 has collapsed for a load lower than 10% of the ultimate load in axial compression, Consequently, a load which is less than, say, 5% of the ultimate load in axial compression can safely be carried on a normally designed structure. To obtain an upper boundary for the load eccentricity below a joint, one can assume the joint to be fixed; the acting moment from the slab can easily be calculated for this assumption. The

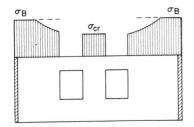

Figure F.17 ☐ The resulting effective buckling stress for a concentrically loaded wall is increased by side restraint. The wall portion between the windows buckles at a stress σ$_{cr}$ while portions closer to the cross walls buckle at higher stresses which, however, cannot exceed the crushing stress σ$_B$.

calculated moment is then divided between the walls above and below the slab end and the load eccentricity is obtained as the moment divided by the axial load in the wall. Since the load and its eccentricity are known, the stresses can be calculated from Eqs. (F.21) and (F.22). The load carrying capacity can be calculated from Eq. (F.27). The risk of buckling can be judged from diagrams in Fig. E.12, and the length factor can be taken as if the structure were fully elastic.

The side restraint of cross walls will increase the buckling load of a wall considerably. The resulting effective buckling stress for a concentrically loaded wall is shown in principle in Fig. F.17. The portion of the wall between the windows is assumed to buckle as a column without side restraints, but the wall portions close to the side walls cannot buckle at all due to the restraint. Therefore the ultimate strength of the masonry will set the limit for the load-carrying capacity in the vicinity of the side wall. The restraining effect diminishes rapidly as the distance from the side wall increases. The effect is well-known from experiments and theoretical calculations for steel plates with axial loads. See, for example, Kollbrunner and Meister [F.12].

References for Chapter F

F.1 ☐ Timoshenko, S.: "Theory of Elastic Stability." Section I.6. McGraw-Hill Book Company, New York, 1961.

F.2 ☐ Sahlin, S.: "Structural Interaction of Walls and Floor Slabs." Inst. för Byggnadsstatik KTH, Meddelande Nr. 33, Stockholm, 1959; Handling No. 35, National Swedish Council for Building Research, Stockholm, 1959.

F.3 ☐ v. Emperger, F. E.: "Versuche mit Eingespannten Balken." Mitteilungen über Versuche. Eisenbeton-Ausschuss des österreichischen Ingenieur-und-Architekten Vereins, Deuticke, Leipzig und Wien, 1913.

F.4 ☐ v. Kazinczy, G.: "Die Bemessung Unvollkommen Eingespannte 'I' Deckenträger unter Berücksichtigung der Plastischen Formänderungen." (The Design of Partially Restrained Steel I-beam Floors Considering Plastic Deformations.) IABSE, Second Volume of the Publications, Zurich, 1933–1934.

F.5 ☐ Sahlin, S.: "Interaction of Masonry Walls and Concrete Slabs," in *Designing, Engineering and Constructing with Masonry Products*, edited by Dr. Franklin Johnson. Copyright © 1969 by Gulf Publishing Company, Houston, Texas. Used by permission.

F.6 ☐ Hellers, B. G., and Sahlin, S.: "Knäcking av tvåledsram med vertikaler utan draghållfasthet." (Buckling of Two-hinged Frame with Verticals without Tensile Strength.) Väg-och Vattenbyggaren Nr. 3, 1966; Bull. No. 60 from Div. Build. Statics and Structural Engineering, Royal Institute of Technology, Stockholm, 1966.

F.7 ☐ Sahlin, S.: "Beräkning av Bärförmågan hos Elementväggar och Murade Väggar." Byggforskningen, Rapport 76, Stockholm, 1962.

F.8 ☐ Nylander, H.: "Undersökning av Bärkraften hos Murade Cementstensväggar." (Investigation of Load-Carrying Capacity of Cement Block Masonry Walls.) Betong, Häfte 3, Stockholm, 1944.

F.9 ☐ Putkonen, A. I.: Vetoa kestämättömiä Pilaria Rasittavan Normalivoiman Epäkeskisyyden Määrittämisestä eräillä Tunnetuilla Kehänlaskumenetelmillä." (Calculation of the Load Eccentricity on Columns Having No Tensile Strength, Application of Some Known Frame Calculation Methods.) Statens Tekniska Forskningsanstalt, Publ. 61, Helsingfors, 1961.

F.10 ☐ Awad, E. Mohamed: "Design Criterion for Masonry Walls Supporting Single-Span Slabs." Term Project for CE 391, University of Illinois, May, 1968.

F.11 ☐ National Institute for Materials Testing, Intyg. U64-3778, Stockholm, 1964.

F.12 ☐ Kollbrunner and Meister: "Ausbeulen." Springer-Verlag, Berlin, 1958.

G □ Masonry walls subjected to inclined forces in their own plane

G.1 □ Introductory remarks

This chapter deals mainly with the ability of masonry walls to withstand horizontal forces (from earthquakes or wind loading) in their own plane. These horizontal forces may be accompanied by vertical forces (dead loads, etc.) of different magnitudes. The strength of a wall loaded in this way is affected by surrounding concrete frames and other confining parts of the building structure, as well as by the strength properties of the wall itself. The comparatively limited information available as a result of different types of tests on masonry is discussed below. Suggested design formulas for the calculation of the failure load for shear walls are also considered.

G.2 □ Inclined, inplane uniaxial load

G.2a: Failure of the mortar bed joints

Compared with the vast number of tests reported on concentrically loaded walls with the load applied vertically, little is known about the strength of masonry walls with the load applied at different inclinations to the horizontal joints. However, some test results are reported. Benjamin and Williams [G.1] carried out tests on shear couplets of two bricks bound together with a mortar joint. Three different mortar types were tested with watered, stiff mud, side-cut, vacuum-treated clay bricks. The test results showed little or no influence of brick and mortar compressive strengths on the couplet bond strengths in tension and shear. A summary of the results of these tests is given in Fig. G.1, where the shear stresses have been calculated as an average stress T/A (shear force divided by wall area) instead of the

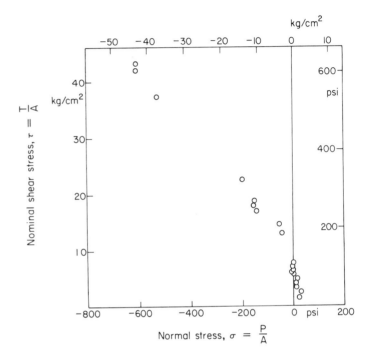

Figure G.1 □ The shear strength τ as a function of the normal stress σ in brick couplets [G.1].

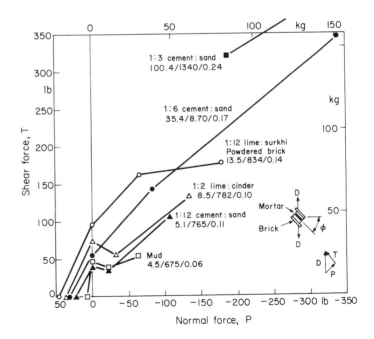

Figure G.2 □ The ultimate shear force T = A · τ as a function of the normal force P = A · σ in mortar joints in brick couplets [G.2]. The different types of mortars employed are indicated on the curves as well as the mortar's tensile strength to compressive strength to modulus of elasticity × 10⁻⁶, all in psi (to obtain kg/cm², divide by 14.22).

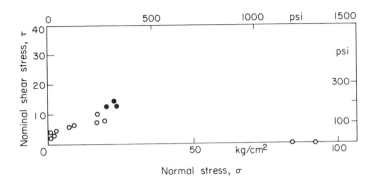

Figure G.4 □ *Shear strength as function of normal stress [G.3]. Compare Fig. G.3. Filled circles indicate bricks of higher strength.*

If the Haller formulas are approximated to a linear relationship, Eq. (G.1), the following equations are obtained:

$$\tau = 160 + 1.0\sigma \text{ psi}$$
$$\sigma \leq 400 \text{ psi}$$

for special quality, and

$$\tau = 50 + 0.88\sigma \text{ psi}$$
$$\sigma \leq 200 \text{ psi}$$

for normal quality.

(G.2c)

The tests of the type shown in Fig. G.3 give the most representative values for shear strengths of the mortar joints in masonry, since disturbances caused by the testing machine platens, etc., are much less likely to occur in this type of test compared with the couplet type tests.

Figure G.5 □ *Shear strength as function of normal stress [G.5].*

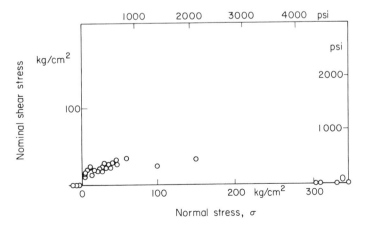

From the limited number of tests mentioned above it seems reasonable to assume that the bond or shear failure of a mortar joint in brick masonry follows Eq. (G.1) in a range of approximately 2 to 15% of the compressive strength of the masonry. τ_B is on the order of 2 to 3% of the compressive strength. However, since the couplet tests [G.1] indicate that the compressive strength of the mortar has no influence on the bond strength, the shear strength should be tested when high stresses are employed. The factor μ was approximately $\frac{1}{2}$ in some masonry tests, near 1 in some other masonry tests, and $\frac{3}{4}$ in the couplet test. In all cases, the stresses are calculated as mean stresses, i.e., the force divided by the area of the joint.

For compressive stresses lower than approximately 2% of the compressive strength of the masonry, and for pure tensile stresses, the shear strength falls below that calculated from Eq. (G.1), as shown in couplet tests and model masonry tests [G.1], [G.5]. The pure tensile bond strength is greatly influenced by workmanship and wetness of the bricks. A suction rate of 20 g/min or less seems to give maximum bond, although saturated bricks produce close to maximum bond [G.1]. It should be mentioned that there are difficulties involved in testing tensile strength, and that the scatter of the results is considerable.

For high compressive stresses the apparent shear strength again is lower than calculated from Eq. (G.1). The few reported tests in this region show failure in the bricks.

Hedstrom [G.6] reports load tests of concrete masonry walls with constant wall dimensions but with the mortar bed joints in 90°, 45°, and 0° inclination to the axial load, which was applied parallel to the longer sides of the walls (Fig. G.6). The obtained ultimate stresses for the three load directions are shown in Fig. G.7 for the two different mortars used. The compressive strengths

Figure G.6 ☐ Block arrangement in walls tested by Hedstrom [G.6] (Fig. G.7).

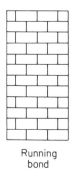

Running
bond

Diagonal
bond

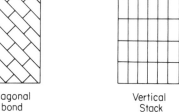

Vertical
Stack

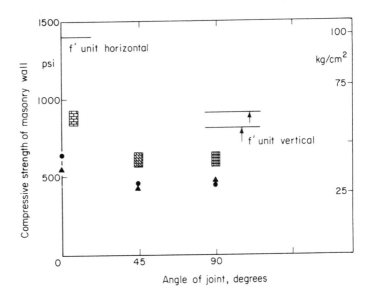

Figure G.7 ☐ Compressive strength of concrete masonry walls tested with the axis of compression inclined against the mortar bed joints [G.6]. Two different mortars were used:

Mix cement:lime:sand	Type of mortar	Mark	Compressive strength (psi)	Tensile strength (psi)	Bond strength (psi)
$1:1\frac{1}{4}:4\frac{3}{4}$	M	●	2000–2500	250–275	58
1:2:7	S	▲	1240–1290	155–180	26

8 × 8 × 16 in. concrete units with 57.1% net area and absorption 15.3 to 17.1% were used.

of the hollow concrete masonry units are also shown for the two principal stress directions (parallel with and perpendicular to the cores). The tensile bond strength (45 and 38 psi) (3.2 and 2.7 kg/cm²) obtained with the two types of mortar was tested on masonry prisms of two blocks in bending. The bond shear strengths and the compressive strengths have been plotted in a diagram (Fig. G.8) of the type shown in Figs. G.4 and G.5. From this diagram the constants τ_B and μ in Eq. (G.1) can be calculated for M mortar to be 3.4 kg/cm² (48.5 psi) and 0.84, respectively, and for S mortar, 1.7 kg/cm² (24 psi) and 0.92, respectively. These figures are supported by too few tests to be conclusive. The shear strength τ_B is of the same order of magnitude as observed on clay brick masonry by Zelger [G.3]

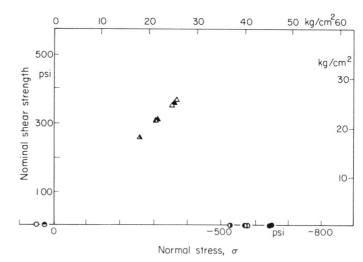

Figure G.8 ☐ *The shear strength as a function of the compressive stress of concrete masonry walls shown in Fig. G.6. The block strength parallel with the holes was about 100 kg/cm² (1400 psi) and perpendicular to the holes 56 to 64 kg/cm² (805 to 915 psi). The values are adjusted to a block strength of 98 kg/cm² (1400 psi).*

Marking	Test type, masonry pattern	Mortar
○	*Bond strength*	*M*
◐	*Bond strength*	*S*
△	*45°, basket weave*	*M*
▲	*45°, running bond*	*M*
◭	*45°, running bond*	*S*
◬	*45°, basket weave*	*S*
●	*Compressive strength, running or stacked*	*M*
◑	*Compressive strength, running or stacked*	*S*
◗	*Compressive strength, running bond*	*S*

($f'_{masonry} \approx 90$ kg/cm²), but much lower than that observed by Yorulmaz and Sozen [G.5] ($f'_{masonry} = 325$ kg/cm²). The "friction coefficient" is higher than observed in all other tests discussed here, except for Haller's [G.13]. The shear strengths τ_B, derived, are 4.5 and 7.8% of the compressive strengths for S and M mortar masonry, respectively, which is somewhat higher than observed on clay brick masonry.

G.2b : Failure of the bricks (and joints)

In tests on masonry with a small angle between the uniaxial load and the vertical line (close to pure compression perpendicular to the joints), Eq. (G.1) no longer applies because of the change in failure mode. Both bricks and joints show cracks at ultimate load and the strength of the bricks comes into play. The transition into the range where bricks start to crack before shear or bond failure in the horizontal joints is reached seems, on the basis of available information, to start at low normal stresses, compared to the strength in direct compression.

In Zelger's test [G.3], two test specimens with normal stresses about 20% of the strength in direct compression were reported to have brick failure; in the Yorulmaz and Sozen test report [G.5], two values of 30 and 45% with brick failures are reported. The question of whether the percentage should be based on the compressive strength is, of course, well founded, but no better experimentally verified base is known to the writer and, furthermore, the compressive strength of the masonry is often taken as a basis for comparisons. However, the tests by Hedstrom [G.6] on concrete block masonry show diagonal strengths of about 70% (66 to 81%) of the strength in direct compression perpendicular to the mortar bed joints, and about the same strength as for compression parallel with the mortar bed joints. Since in Zelger's test the bricks had 28.5% of the compressive strength parallel to the cores when loaded perpendicular to the cores, and brick failures appeared in the masonry at 20% of the direct compressive strength of the masonry, a safe rule of thumb could be that Eq. (G.1) applies only up to the least of the masonry's compressive strengths horizontally or vertically. From there on the masonry should only be loaded up to the "minimum" compressive strength when the load is inclined.

G.3 □ Biaxial load

Few biaxial tests on masonry have been run; a few are reported by Hedstrom, Litvin, and Hanson [G.12]. It is possible that the strength of biaxially loaded masonry is different from the strength of uniaxially loaded masonry, and thus the internal-friction theory has to be modified. For concrete it is known that the compressive strength in one direction is to some extent affected by the stress in a perpendicular direction; see Hilsdorf [G.7]. This is particularly true if the perpendicular stress is tension, and this fact is an indication that the shear strength in a joint can depend upon not only the normal stress, but also on the stress parallel to the joint.

G.4 □ Shear walls

G.4a: Racking test

The present standard test in the United States for estimating the shear strength of a masonry wall is the racking test, described in ASTM E72. The boundary conditions and the stresses produced in the test walls by this method are unclear (see Fig. G.9), and therefore only a relative measure of the racking or shearing strength of the wall specimen is obtained from such tests. The obtained apparent shear strength as calculated from

$$\tau = \frac{T}{ld} \tag{G.4}$$

is usually of the order 2 to 20 kg/cm² (25 to 300 psi) under laboratory conditions. See [G.8] and [G.9].

These racking strength stresses correspond to 5 to 9% of the compressive strength of similar walls (walls with structural clay tile loaded sidewide excluded) [G.8], [G.9].

For concrete masonry walls, the racking strength (Table G.1) was reported by Fishburn [G.10] to be 1.8 to 3.4 kg/cm² (25 to 50 psi) for masonry walls having a compressive strength of 27.5 to 33 kg/cm² (390 to 470 psi), giving a racking strength of about 7 to 10% of the compressive strength.

Figure G.9 □ Arrangement of racking tests and the force distribution on the test specimen.

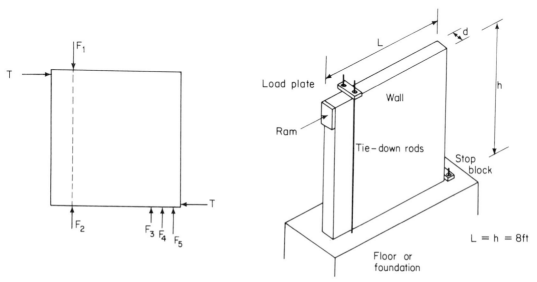

Table G.1 □ *Racking strength of concrete masonry walls, made of concrete masonry blocks having a compressive strength of 80* kg/cm² *(1140* psi) *gross area or 150* kg/cm² *(2150* psi) *net area* [G.10].

Type of mortar	Mortar		Masonry racking strength [psi] (kg/cm²)
	Compressive strength [psi] (kg/cm²)	Bond strength [psi] (kg/cm²)	
(*ASTM*) *N* (masonry cement *B* –sand)	[740] (52)	[9] (0.63)	[26.2] (1.82)
N (masonry cement *C*-sand)	[890] (62)	[11] (0.77)	[39.3] (2.73)
S (masonry cement *B*-Portland cement-sand)	[1750] (122)	[20] (1.4)	[43.6] (3.04)
S (masonry cement *C*-Portland cement-sand)	[2060] (144)	[23] (1.6)	[49.1] (3.42)

G.4b: Horizontally and vertically loaded wall without frame

The load-carrying capacity of a wall subjected to a horizontal force at one of the upper corners is governed mainly by the shear and tensile strength of the bed joints at the foundation of the wall. By precompression, for example by dead load from slabs and walls above, the strength is increased in a manner similar to that described for masonry specimens loaded with an inclined load.

Murthy and Hendry [G.11] report "$\frac{1}{6}$-model" tests† on three-bay, one-story shear walls 0.669 in. thick about 16 × 16 in. in height and length. The bricks had an average apparent strength of 311 kg/cm² (4421 psi), and the cement–sand mortar about 85 kg/cm² (1200 psi). The horizontal shear strength was tested for various additional vertical loads up to 12.7 kg/cm² (180 psi) and the following relationship was established:

$$\tau = 2 + 0.5\sigma \text{ kg/cm}^2 \tag{G.5}$$

$$\sigma \leq 12.7$$

$$\tau = 30 + 0.5\sigma \text{ psi} \tag{G.6}$$

$$\sigma \leq 180$$

with slightly higher values for $\sigma < 5.3$ kg/cm² (75 psi) and lower values for $\sigma \geq 7.7$ kg/cm² (110 psi); see Fig. G.10. The plateau of the shear stresses of about 5 kg/cm² (70 psi) is explained by Murthy and Hendry as a breakdown of the bond between the mortar and the bricks.

†Professor Hendry uses the expression "$\frac{1}{6}$-model" to mean that the model is scaled to one-sixth of full size.

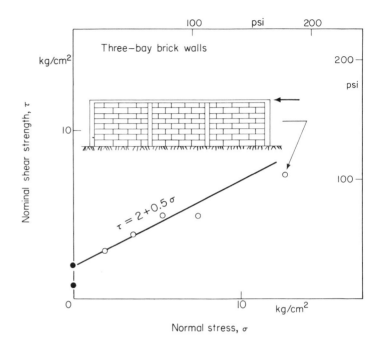

Figure G.10 □ The shear strength as a function of the normal stress in a masonry wall. Tests on three-bay model brick masonry walls [G.11] are marked ○, and tests on one-bay brick masonry wall [G.1] are marked ●.

Benjamin and Williams [G.1] tested model walls without frames and without vertical loads and found apparent shear strengths of 1 to 2 kg/cm² (15 to 30 psi); see the filled circles in Fig. G.10.

The wall tests mentioned above gave shear strengths as a function of precompression which are in good agreement with the tests by Zelger [G.3], shown in Fig. G.4, but are considerably lower than the values obtained by Yorulmaz and Sozen on masonry units, Fig. G.5, and by Benjamin and Williams as well as by Krishna and Chandra on brick couplets, Figs. G.1 and G.2. The values are also lower than those observed by Haller [G.13], Eq. (G.2c). The brick couplets probably give higher strengths because of different loading type and stress distribution. The high values obtained by Yorulmaz and Sozen could probably be accounted for by the high brick and mortar strengths, although Murthy and Hendry used a mortar with about the same strength but weaker bricks. It therefore seems wise not to exceed the values calculated from Eqs. (G.5) and (G.6) without compelling reasons.

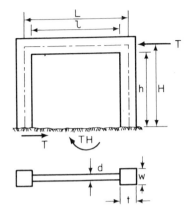

Figure G.11 □ Type of test specimen employed for the test results plotted in Fig. G.12 [G.1].

G.4c : Wall with concrete frame

G.4c1 : Shear strength of walls with strong frames. The test results obtained by Benjamin and Williams [G.1] on frames with masonry shear walls (Fig. G.11) have been compiled in Fig. G.12. The tests were run with different sizes and types of frames as well as without frames. The tests without frames gave very low strength, which could be calculated on the basis of failure in tension (bending) between the wall and the foundation. The load-carrying capacities obtained on solid walls with frames strong enough to withstand the overturning moment are closely correlated with the horizontal mortar joint area, i.e., the wall length times the wall thickness. The following equations give a good approximation within the range of the test:

$$\frac{T}{A} = 150 \text{ psi} \tag{G.7}$$

$$\frac{T}{A} = 10.5 \text{ kg/cm}^2 \tag{G.8}$$

with the horizontal force T and the area A. It can also be noted from these tests that five seemingly identical model frames were tested and the test results showed a coefficient of variation of 18%, while the coefficient of variation for the test specimens having varying reinforcement and concrete area was slightly less, the specimens were taken regardless of frame dimensions but with the same wall section. In the test reported, the frames were designed to be stronger in overturning than the walls in shear.

Figure G.12 □ Ultimate horizontal force T as a function of horizontal wall area A. Frames without vertical load [G.1], Fig. G.11. Different wall scales were used: ● = 0.34, △ = 0.25, ○ □ ■ = 0.42, *and* ▲ = 1.0.

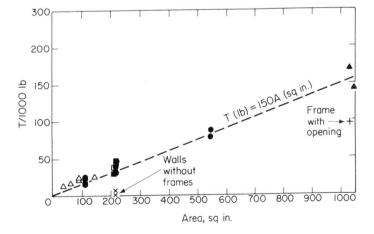

G.4c2:   Influence of reinforcement on cracking.   Yorulmaz and Sozen [G.5] tested model frames in scale 1:8 in
order to study the effect to reinforcement on the strength of a
reinforced concrete frame with filler walls of model brick
masonry. The Benjamin and Williams test [G.1] also included
three test specimens to study this effect. With the assumption
that the forces increase by the second power of the scale of the
test specimens, a diagram showing the influence of reinforcement in the frame columns on the load-carrying capacity of a
frame has been complied (Fig. G.13). Yorulmaz and Sozen also
report the cracking load which on some occasions (little reinforcement) is the same as the failure load.

The failure load increases with increasing percentage of reinforcement (0.7 to 2.2) in the Yorulmaz–Sozen test (7 tests), but
no clear tendency can be seen for the cracking load or the failure
load (3 tests) obtained in the Benjamin–Williams tests. A direct
comparison of the absolute values of the failure loads from the
two sources cannot be made, since the length over height ratios
are different (2.0 and 1.74), and so are the length over thickness
ratios (34 and 15.5).

For frames with very low strengths, it is probable that the
rather high shear stress of 10.5 kg/cm² (150 psi) in Eq. (G.7)

Figure G.13 □ *Ultimate horizontal*
force and cracking load H plotted
against percentage reinforcement in
the concrete frame around a
masonry wall.
▲ *is ultimate load [G.5]*
○ *is cracking load [G.5]*
● *is ultimate load [G.1]*

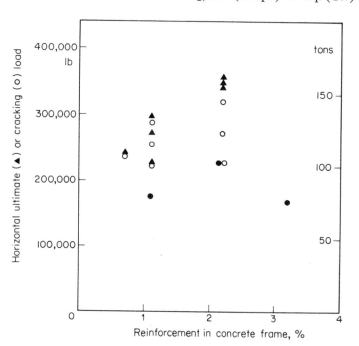

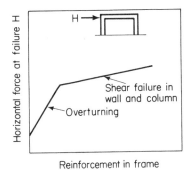

Figure G.14 □ Estimated principal relationship between percentage of reinforcement in concrete frame and horizontal load H at failure for concrete frame with masonry filler wall.

decreases, and for no reinforcement or frame at all the shear stress should reach the value predicted by Eqs. (G.5) and (G.6). The principal effect of reinforcement in the frame could consequently be represented by two lines in the diagram shown in Fig. G.14. Some simple calculations indicate that the breaking point should be around 0.1 to 0.5% of reinforcement.

G.4c3: *Deflections.* A horizontal deflection formula for the lateral displacement of a brick wall in concrete frame was proposed by Benjamin and Williams [G.1]:

$$\delta = \frac{1.2hT}{ldG} \tag{G.9}$$

where δ is the wall deflection, T is the wall shear, h is the panel height, l is the panel length, d is the panel thickness, and G is the shearing modulus of brick composite (35,000 kg/cm² or 500,000 psi, approximately). This formula predicts the deflections in the elastic or precracking stage only. The tests showed postcracking deflections several times larger than the elastic ones.

G.4c4: *Postcracking deflection and energy absorption.* The deflections and the development of a final frame failure after cracking of the wall have been calculated by Yorulmaz and Sozen [G.5] for the frames they tested. The modes of failure discussed are the following (Fig. G.15):

1. Cracking and yielding of tension column.
2. Cracking in the wall → axial tension in the beam and shear and axial tension in the tension column → shear failure in the tension column, eventually after beam cracking → loads to the compression column which yields or shears off.
3. Cracking in the wall → axial tension in the beam and shear and axial tension in the tension column → crack in top beam → yielding of the beam → load to the compression column which yields or shears off.

With these assumptions, the load–deflection curves have been calculated to very large deflections, of the order of one-third of the dimension of the column. After the collapse mechanism 2 has developed, the wall might still carry load due to the friction in the cracks. The maximum load-carrying capacity of this mechanism (bottom of Fig. G.15) was found to be approximately

$$P = \frac{A_s \sigma_y}{2\sqrt{2}} + \frac{A_s \sigma_y (\sqrt{2} + 1)}{2\sqrt{2}} \mu \tag{G.10}$$

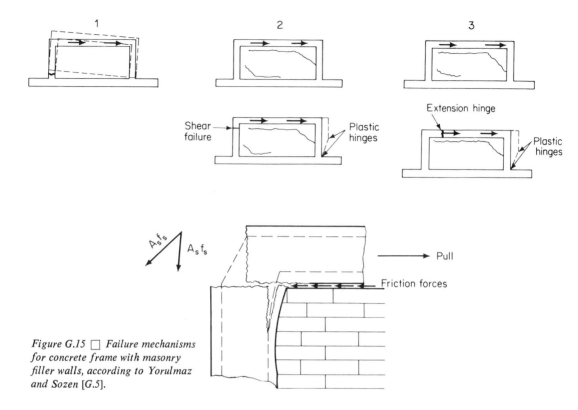

Figure G.15 □ Failure mechanisms for concrete frame with masonry filler walls, according to Yorulmaz and Sozen [G.5].

where μ is the friction coefficient between the wall and the concrete and could be taken as 0.45. Thus

$$P = 0.75A_s\sigma_y \qquad (G.11)$$

With the considerations mentioned above Yorulmaz and Sozen calculated deflections and established safe lower bounds for the energy-absorbing capacity for a one-story concrete frame with filler in walls of brick masonry.

G.4c5: An analysis of reinforced concrete frames with filler walls.   Feodorkiw [G.14] used a lumped-parameter model to represent a reinforced concrete frame with masonry filler walls subjected to inplane forces. The model's response to loads was then programmed and calculated numerically on an IBM 7094–1401 system. The progressive locations of cracking within the structure were determined on the basis of successive solutions as load was increased up to ultimate. The calculations show that the same ultimate load capacity may be generally expected, irrespective of the value of filler modulus, provided shear

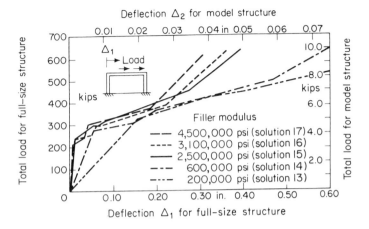

Figure G.16 □ Computed load-deflection plots for single-story structures. p = 2.2%, modulus of elasticity of frame = 285,000 kg/cm² (2 × 10⁶ psi) [G.14].

failure in the tension column is prevented. This is of particular significance in cases where a filler has a very low elastic modulus as a result of poor workmanship or low quality materials.

For filler with an elastic modulus less than about ¼ of the frames, the deformations are much larger than for stiffer filler, even in the precracking stage (Fig. G.16). An empty frame has still higher deformations but considerably lower load-carrying capacity. Thus, a poor filler contributes much to the strength of a frame, even though it needs a certain stiffness to drastically reduce the deformations. Beyond a certain point, increased filler stiffness does not reduce the deformations at the same high rate as for lower stiffnesses (Fig. G.17).

Figure G.17 □ Variation of deflection $\triangle_1$ with elastic modulus of filler. Elastic modulus of frame was assumed to be 285,000 kg/cm² (2 × 10⁶ psi), theoretical values [G.14].

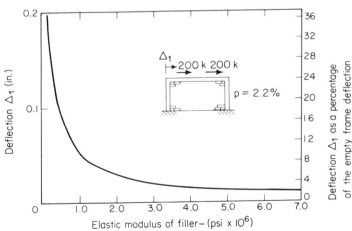

Variation of deflection Δ_1 with elastic modulus of filler

G.5 ☐ Summary

For a region of low normal stresses ($<15\%$ of the compressive strength), tests show a relationship between the failure shear stress in a mortar joint and the normal stress of the following type:

$$\tau = \tau_B + \mu\sigma \qquad (G.1)$$

where $\tau_B \approx 2\,\text{kg/cm}^2$ (30 psi) and $\mu \sim 0.5$ for some reported qualities of masonry. This equation seems to be invalid for very low normal compressive stresses and normal tensile stresses, where the observed shear stresses are lower (see Figs. G.1 and G.5). For higher normal stresses the failure can eventually take place in the bricks. Considerable higher ultimate shear stresses have been observed in favorable cases.

The stress distribution in a shear wall might be calculated on a basis of elasticity with more or less rough approximations; see Krishna and Chandra [G.2]. Yorulmaz and Sozen [G.5], and Feodorkiw [G.14]. The deflections of a single uncracked shear wall in a concrete frame might be calculated from

$$\delta = \frac{1.2hT}{ldG} \qquad (G.9)$$

The cracking load for horizontal forces can be estimated from the equation

$$T = \tau ld \qquad (G.4)$$

where τ is about $10\,\text{kg/cm}^2$ (150 psi) with excellent workmanship. The postcracking behavior must be calculated under consideration of several different failure mechanisms, and the final load-carrying capacity according to Yorulmaz and Sozen [G.5] is

$$P = 0.75A_s\sigma_y \qquad (G.11)$$

for the arrangement of reinforcement shown in Fig. G.15.

The dynamic damping factor in the uncracked stage varies with the applied strain, see Krishna and Chandra [G.2]:

$$f \sim 2 + \frac{\varepsilon}{8}\cdot 10^{-3}\% \qquad (G.12)$$

The most important factors in obtaining a high shear resistance in bond or shear in masonry are suction rate, workmanship, and mortar strength.

A great number of analytical methods for calculation of shear walls are dealt with at length in the published Proceedings of

H □ Bending and combined axial and lateral load on masonry walls

H.1 □ Introductory remarks

Masonry walls are sometimes loaded mainly with lateral loads and sometimes span only in a vertical direction. Consider, for example, a piece of wall between two openings (Fig. H.1). If this wall does not carry any load in the vertical direction (the slabs may span parallel to the wall), the wall would be subjected to pure bending by, for example, wind loading. Cracking and failure would then occur as soon as the tensile strength in bending of the wall were reached if no vertical load could be mobilized. Thus, the modulus of rupture of the masonry is of interest to know. In the following, a number of factors affecting the modulus of rupture are studied.

In Section H.3 the case of bending of a wall supported only along the vertical sides is studied; in Sections H.6, H.7, and H.8, the combination of axial and lateral load is discussed. Arching action and walls supported along four sides are also discussed at the end of this chapter.

Figure H.1 □ Part of a building showing a masonry wall spanning in a vertical direction only.

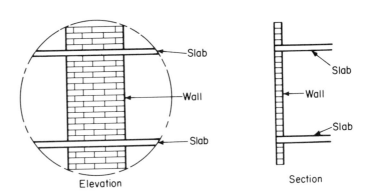

137

H.2 ☐ Bending of a masonry wall spanning vertically: modulus of rupture

H.2a: General

Consider the wall shown in Fig. H.1. This wall is assumed to be subjected to pure bending by lateral forces. The modulus of rupture of the masonry will, therefore, determine the load-carrying capacity against the lateral loads. The various factors affecting the modulus of rupture are discussed below.

H.2b: Effect of mortar tensile strength and bond

Since a masonry wall spanning vertically fails in the mortar joint (tensile failure in mortar) or in the interface between the mortar bed joint and the block or brick (bond failure) or sometimes partly in the mortar, partly at the interface, and partly in the brick or block (mortar, bond, brick failure), it is expected that the mortar tensile strength and the bond strength are important factors for the modulus of rupture of masonry.

The nominal modulus of rupture based on the gross area for concrete block masonry walls was reported by Hedstrom [H.1]

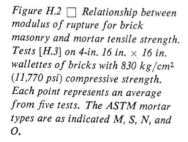

Figure H.2 ☐ Relationship between modulus of rupture for brick masonry and mortar tensile strength. Tests [H.3] on 4-in. 16 in. × 16 in. wallettes of bricks with 830 kg/cm² (11,770 psi) compressive strength. Each point represents an average from five tests. The ASTM mortar types are as indicated M, S, N, and O.

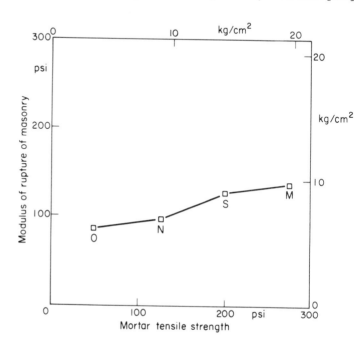

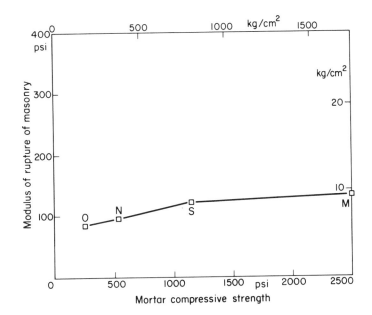

Figure H.3 ☐ Relationship between modulus of rupture and mortar compressive strength for brick masonry wallettes 4-in. thick, 16 in. × 16 in. built of bricks with 830 kg/cm² (11,770 psi) compressive strength. Each point represents an average from five tests. Compare Fig. H.2 [H.3].

to be about 3.5 kg/cm² (25 psi) for walls in S mortar and about 7 kg/cm² (50 psi) for M mortar, the values slightly higher for stacked bond and slightly lower for running bond. (The vertical continuous mortar joints in stack bond apparently acted like a thin mortar beam.)

Fishburn [H.2] found that the modulus of rupture of concrete masonry walls (in the range 23 to 53 psi) was about 3% of the mortar compressive strength for N mortar (47 to 83 kg/cm², 670 to 1180 psi) and about 1.8% for S mortar (126 to 2040 kg/cm², 1800 to 3030 psi). The modulus of rupture of the concrete masonry wall was approximately 65% of the bond strength found on couplets.

Tests (SCPRF [H.3]) show that the tensile strength of the mortar influenced the modulus of rupture of brick masonry but not in direct proportion (Fig. H.2) to the tensile strength. Since the tensile strength of mortar varies with the compressive strength of the mortar, the modulus of rupture of the brick masonry also varies with the compressive strength of the mortar (Fig. H.3). The primary factor, however, is probably the tensile and bond strength of the mortar.

Since the ratio between the modulus of rupture of brick masonry and the mortar tensile strength is not even nearly constant for different values of the mortar tensile strength, there must be also other important factors changing with the mortar tensile and bond strength. It is said (Benjamin and Williams [H.4]) that the bond strength between the mortar and the bricks is best indicated by flow and water retentivity tests. The water to cement ratio is also, according to the same authors, an important variable; however, the authors do not report any data.

Some data on the influence of suction (IRA) on the modulus of rupture are reported in what follows.

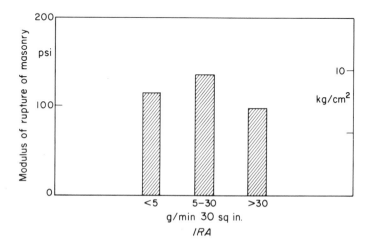

Figure H.4 □ Relationship between modulus of rupture of masonry and initial rate of absorption (IRA) of the bricks. S mortar and $\frac{3}{8}$-in. mortar joints [H.3].

Figure H.5 □ Relationship between tensile bond strength of masonry and suction of bricks. Two different mortars with the cement to lime to sand ratios indicated [H.5].

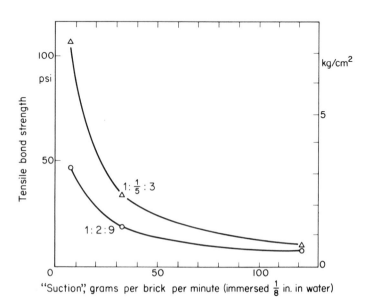

H.2c:　Effect of suction (IRA) and flow

It is known that high suction or IRA of the bricks or blocks is detrimental to most strength properties of the masonry. The modulus of rupture of brick masonry was tested at SCPRF [H.3] for different bricks with different suctions but with the same mortar. The results are summarized in Fig. H.4. As seen from the figure, the highest modulus of rupture is obtained for a suction of 5 to 30 g/min. The decrease in modulus for high suction probably does not show up in full magnitude in these tests since bricks with suction (IRA) higher than 20 g/min per 30 sq in. were wetted before being laid. The tensile bond strength decreases rapidly with high suction of the bricks, as can be seen from Fig. H.5, taken from Plummer and Reardon [H.5]. The strongest mortar with most cement content was most affected.

However, the influence of the suction on the tensile strength of masonry varies not only with the mortar strength (or cement content) but also with the flow of the mortar, as can be seen in Fig. H.6, derived by Plummer and Reardon [H.5]. This figure must still be read with caution since other mortar properties were not held constant for all the points indicated in the diagram.

Figure H.6 □ *Relationship between masonry tensile bond strength and "suction" for mortars of different flows (mortars of same flow may have differences in other properties) [H.5].*

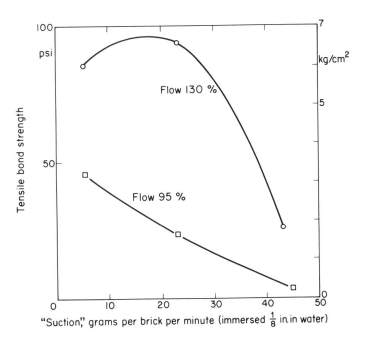

To summarize, the modulus of rupture depends upon mortar strength, mortar flow, water to cement ratio, retentivity, brick IRA (suction), and probably also on the absorption and the curing conditions.

H.2d: Effect of joint thickness

According to test results published by SCPRF [H.3], the modulus of rupture of masonry decreases as the joint thickness increases. See Fig. H.7. In this case the modulus of rupture decreased from 10.8 kg/cm² (154 psi) to 4.5 kg/cm² (64 psi) when the mortar joint thickness increased from 0.62 to 1.9 cm ($\frac{1}{4}$ to $\frac{3}{4}$ in.). Furthermore, the modulus of rupture is considerably less than the tensile strength of the mortar tested on prisms, which, of course, can be due to different curing conditions.

The inverse variation of modulus of rupture with the joint thicknesses cannot easily be allocated to one single variable,

Figure H.7 ☐ Relationship between modulus of rupture of masonry and mortar bed joint thickness. Four-in. wallettes 16 in. × 16 in. bricks with 830 kg/cm² (11,770 psi) compressive strength, Mortar type S [H.3].

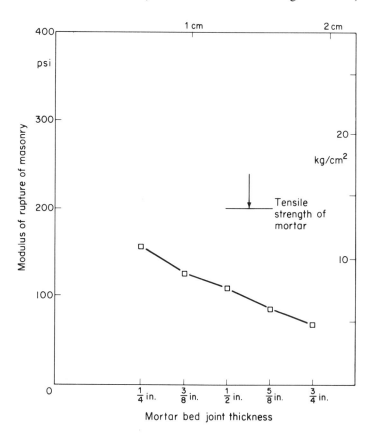

especially since the type of failure is not known. The curing conditions vary with the joint thickness, however, and the workmanship could also be influenced by a change in thickness. When the joint thickness increases, the chances that a weak spot will be enclosed in the joint also increase, and thus the statistical effect could be a part of the cause.

H.2e : Effect of brick strength

The failure of masonry in bending often takes place in the mortar or in the interface between mortar and brick. Thus, it would be reasonable to assume that the strength of the bricks has little effect on the modulus of rupture of the masonry. Yet tests at SCPRF [H.3] and by Davis (reported by Plummer and Reardon, [H.5], p. 71) show a strong influence of the brick strength on

Figure H.8 ☐ *Relationship between modulus of rupture of masonry and brick strength*
☐ *is 4-in. walls 4 × 8 ft built with S mortar $\frac{3}{8}$-in. joints [H.3]*
■ *is 8-in. walls 4 × 8 ft built with S mortar $\frac{3}{8}$-in. joints [H.3]*
○ *is 5-in. walls 4 × 8 ft built with S mortar $\frac{3}{8}$-in. joints [H.28].*
▲ *is walls with 1:1:6 mortar [H.5]*
△ *is walls with 1:$\frac{1}{2}$:4$\frac{1}{2}$ mortar [H.5]*

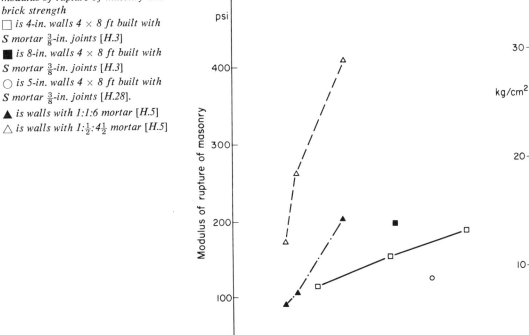

the modulus of rupture of the masonry (see Fig. H.8). The relationship, however, is different for different mortar mixes, and probably different for different wall thicknesses. The modulus of rupture is influenced by the brick strength because of the change in IRA which usually follows the change in brick strength. This latter assumption is supported by tests reported by Hallquist [H.6]. He reports bending tests on 10-brick columns built with lime–cement–sand mortar KC 35/65 (weight proportions of lime to mortar, 35 to 65) and with two types of bricks, one medium and one hard-burned. The modulus of rupture was 2.3 kg/cm² (33 psi) for the medium-burned (high suction) and 7.38 kg/cm² (103 psi) for the hard-burned (low suction) brick wall. Some test specimens built of medium-burned bricks containing salts giving an almost glassy brick surface (extremely low suction) had too low a modulus of rupture to allow transporting to the testing machine. Thus, also, the decrease in strength properties for very low suction (Fig. H.4) was confirmed by these tests. Finally, there is a possibility that very low strength bricks fail in the extreme surface layer.

H.2f :   Effect of variation in masonry compressive strength

Since the compressive strength of masonry is influenced by many of the factors which determine the modulus of rupture of the masonry, it is conceivable that a positive correlation between the

Figure H.9 □ Relationship between modulus of rupture of masonry and masonry strength in compression
□ *is 4-in. walls, S mortar [H.3]*
○ *is 5-in. walls [H.3]*
▲ *is 8-in. walls [H.3]*

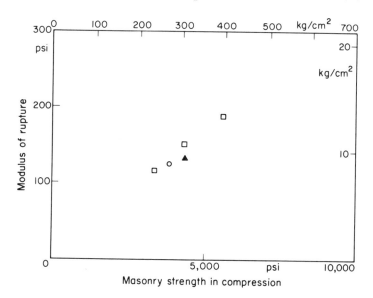

modulus of rupture and the compressive strength of masonry
should exist. A relationship of this kind can be seen in Fig. H.9,
which is compiled from data reported by SCPRF [H.3]. The
modulus of rupture is in this case about 4% of the compressive
strength. The percentage was higher in some tests on concrete
block masonry walls reported by Hedstrom [H.1]: 4.5 for *S*
mortar and 6.3 for *M* mortar, based on gross area with blocks
having absorptions of 15.3 to 17.1% and 1400 psi compressive
strength. Tests by Fishburn [H.2] gave about 3% for *N* mortar
and 4 to 5% for *S* mortar. Since so many factors are involved
and the number of tests are few, these figures should be used
with caution. Direct tests on actual material are much to be
preferred.

H.3 □ Bending of a masonry wall spanning horizontally

This loading case would occur if a masonry wall were completely free at its upper and lower edges but supported by columns at the vertical edges and the load, such as wind, acted
laterally on the wall.

In tests on horizontally spanning concrete block masonry walls,
Cox and Ennenga [H.7] measured a modulus of rupture of about
7.8 kg/cm² (110 psi) calculated on the gross area. The walls
were made of $8 \times 8 \times 16$ in. hollow concrete blocks. For
cavity walls of two wythes of blocks $4 \times 8 \times 16$ in. (two single
wythes each 4 in. thick, with a cavity between them), the observed modulus of rupture, based on gross area, was about 11.2
kg/cm² (160 psi), if it is assumed that the applied moment is
divided equally between the two wythes. The walls were laid in
mortar of two parts Portland cement, one part masonry cement,
and three parts sand.

Hedstrom [H.1] reports tensile strengths of concrete masonry
walls of about 7 kg/cm² (100 psi) based on the net area, a value
which compares reasonably well to Cox and Ennenga's values
which, however, should be about 50% higher on a net area
basis; but due to the test setup it is possible that Hedstrom's observed values were somewhat too high. Hedstrom has also
reported tests on transversely loaded concrete masonry walls
[H.1] which in a running bond showed a nominal modulus of
rupture based on the gross area of about 5.6 kg/cm² (80 psi) or
8.4 kg/cm² (120 psi) based on net area.

A concrete block masonry wall in running bond laid of $8 \times
8 \times 16$ in. hollow units in *M* or *S* mortar will therefore normal-

ly develop a modulus of rupture of 5 to 8 kg/cm² (70 to 110 psi) based on the gross area. The tensile strength testing by Hedstrom [H.1] indicates that the modulus of rupture (for strong mortars giving a crack through the blocks instead of a zigzag crack following joints) can be estimated by taking the concrete's tensile strength as 8% of the concrete's compressive strength (net areas), and the mortar's tensile strength as about 7 to 10% of the compressive strength for M and S mortars and 15 to 20% for N and O mortars, and calculating the average tensile strength along a vertical, straight crack through the blocks and joints. This method gives obtained values which are somewhat too high, because the stress concentrations in the blocks at the vertical joint ends are not taken into account. Therefore, a high factor of safety should be employed.

Nilsson [H.23] observed a modulus of rupture in horizontal bending that was three to six times as high as the modulus of rupture in vertical bending (with the failure in one bed joint). The tests were run on six different brick makes using two different mortar mixes. The test of the modulus of rupture of brick masonry walls reported by Hallquist [H.6] was accompanied by tests on brick masonry specimens bent in a "horizontal" direction, and the specimens having very low modulus of rupture in "vertical" direction showed a modulus of rupture of 11.4 kg/cm² (160 psi) when spanning "horizontally." It can thus be concluded from these tests that the modulus of rupture is different in different directions of the masonry, especially for low strength mortars or poor bonds. The main reason for the difference is the interlocking of the blocks and the friction between them when the wall spans horizontally. The mechanism was described by Royen [H.29] and is briefly outlined in what follows.

If one assumes that the shear stress over the area ($d \times z$) in Fig. H.10 never exceeds τ (= the yield stress in shear = the friction stress), the maximum moment carried by one such area is

$$M = \frac{\tau d^2}{2}\left(z - \frac{d}{3}\right) \tag{H.1}$$

which can be derived by the aid of the membrane analogy (see Timoshenko [H.8]). The number of such "shear areas" over the wall height h is approximately h/b, and therefore the moment capacity per unit height of the wall is

$$M = \frac{\tau d^2}{2b}\left(z - \frac{d}{3}\right) \qquad z \geq d, l \geq 2z \tag{H.2}$$

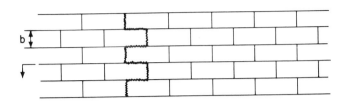

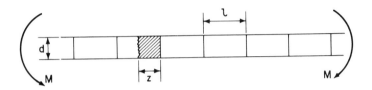

Figure H.10 □ Mechanism of failure for masonry spanning horizontally. The area on which friction acts is z × d.

If the shear stress τ is mobilized solely by friction, the vertical stress σ times the friction coefficient can be substituted for τ:

$$M = \frac{\mu \sigma d^2}{2b}\left(z - \frac{d}{3}\right) \tag{H.3}$$

According to Royen's original formula, consideration was also given to the uneven distribution of normal stress over the area of friction, so that when the axial force was applied outside the kern, the friction area was reduced from $d \times z$ to $\frac{3}{2}(d - 2e) \times z$, and furthermore the stress was taken as an average over the remaining area.

An upper limit to the obtainable moment according to Eq. (H.3) is the moment which causes a bending failure with cracks through the bricks and the vertical joints.

$$M = \frac{b^2 \sigma_t}{6} \tag{H.4}$$

The coefficient of friction for light weight cellular concrete was approximately 0.8 [H.9] and for clay brick masonry and concrete block masonry about 0.5 to 0.7 in shear (see Chapter G); thus

$$M \approx \frac{\mu \sigma d^2}{2b}\left(z - \frac{d}{3}\right) \leq \frac{b^2 \sigma_t}{6} \qquad z \geq d, l \geq 2z, \tag{H.5}$$
$$\mu \approx 0.5 \text{ to } 0.8$$

For masonry such as Siporex or Ytong having no mortar in the joints, but rather joined by tongue and groove or by small plastic discs, the shear load capacity of the connectors should be added [H.9].

H.4 ☐ Modulus of elasticity in bending

Different moduli of elasticity must be expected for different cases of loading, since the strength, thickness, confinement, state of carbonation, etc., in the mortar joints varies over the wall thickness. When the axis of bending or thrust is inclined with respect to the bed joints, the interlocking effects also influence the magnitude of the modulus of elasticity.

The modulus of elasticity, observed in tests run by SCPRF [H.3], was slightly higher in bending than in compression for masonry of high strength bricks, and about the same in both

Figure H.11 ☐ Relationship between modulus of elasticity of masonry in bending and compression and brick strength, S mortar. For some tests the masonry strength in compression is indicated [H.3], [H.27], [H.28].
☐ is bending 4-in. walls
○ is compression 4-in. walls
● is 5-in. walls
■ is 8-in. walls
C is compression
B is bending
M is metal ties

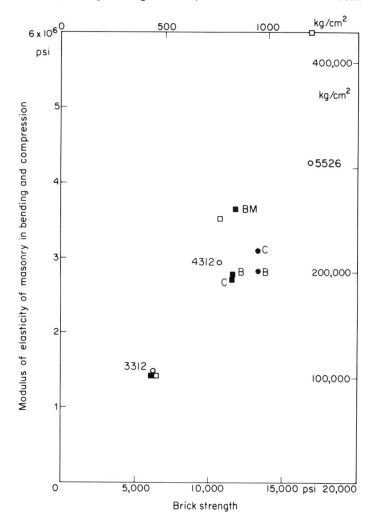

cases for masonry of medium strength bricks, all of type S mortar (Fig. H.11). (The compressive strength of the masonry was of course also affected by the changes in brick strength in the tests, as indicated in Fig. H.11.) It is easily conceived that stress distributions in the bed joints could exist, giving lower modulus of elasticity in bending than in direct compression.

The tests by Hallquist [H.6] mentioned earlier showed a difference in modulus of elasticity in bending vertically and horizontally. The "vertical" bending gave a modulus of elasticity of 50,000 kg/cm² (710,000 psi), while the bending in "horizontal" direction indicated a modulus of elasticity of 87,000 kg/cm² (1,240,000 psi); i.e., the effect of interlocking and the smaller number of mortar joints (Section D.2a) in "horizontal" direction increased the modulus of elasticity by about 75% in that particular case.

H.5 □ Inclined axis of bending

The available information about the strength of masonry when the axis of bending is oblique to the joints is limited. However, Hedstrom [H.1] reports tests on patterned concrete masonry walls with some of the test results from bending of walls with running bond or stack bond plotted in Fig. H.12. The figure supports the assumption that the bending moment resistance is gradually increased with increasing angle between the bed joints and the axis of bending up to 90°. The resistance is, for 90° rotation, increased by about 100% for walls of M mortar and 300% for walls of S mortar. For stack-bonded walls, a decrease in moment capacity has been found, as may be noted in Fig. H.12.

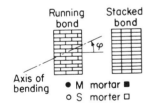

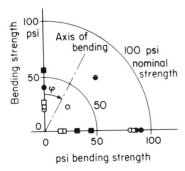

Figure H.12 □ *Moment capacity of masonry as a function of the direction of the bending moment [H.1]. Hollow concrete blocks 8 × 8 × 16 in. block strength about 100 kg/cm² (1400 psi).*

Strength of Masonry	Mortar	
	M	S
$f'_{vert} \sim$	640 psi	580
$f'_{hor} \sim$	440	480

H.6 □ Combined axial and lateral load

H.6a : Introductory remarks

When a wall is not only bent by a lateral load but also compressed by an axial load, the simultaneous action of the loads must be taken into account. The axial load acts in a manner somewhat similar to a reinforcement or a prestressing force under certain conditions. On the other hand, if the axial load is eccentrically applied and gives a moment acting in the same direction as the lateral load, the wall resistance against lateral load is sometimes reduced by the axial load. If wall movements in the wall plane are prevented at the supports, an arching action can take place in the wall, greatly improving the wall resistance against lateral loads. If the wall is spanning both horizontally and vertically, the load-carrying capacity of the wall is improved by "slab" action. These phenomena are discussed in greater detail in the following sections of this chapter.

H.6b : Concentric axial load and lateral loads

Bending tests by Hedstrom [H.1] on precompressed and axially loaded free concrete masonry walls showed that the cracking load for the wall could be calculated from standard flexural formulas into which the known value of tensile bond strength was inserted. After the first cracking, the precompressed walls continued to carry increasing lateral loads up to about 1.25 times the cracking load. The reason for the ability of the axially compressed wall to carry loads higher than the cracking load (this was not the case for the walls loaded laterally only) is discussed theoretically in the following.

With the assumption that the material has no tensile strength and follows Hooke's law on compression, the ultimate lateral concentrated loads have been calculated [H.10], [H.11]. The basic equations for the calculations are the same as in Chapter E, but part of the solution must be graphical.

The wall is assumed to be split at the loading points into three portions for which the solutions are known. The two end pieces have then the loading condition described for the wall in Fig. E.8. The middle piece is axially loaded with the same eccentricity at both ends. The solutions for the three pieces are combined by trial and error under the conditions that the equations of equilibrium are satisfied and that the deformations are compatible at the joining points.

Figure H.13 □ Relationship between the transverse load H and the compressive load for compression members under the conditions of loading shown. The ordinate represents Hh/Pd for the full-line curve and $Hh/P_E d$ for the dash-line curve.

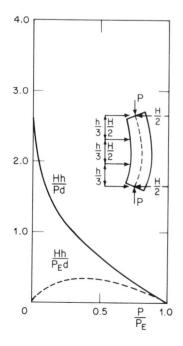

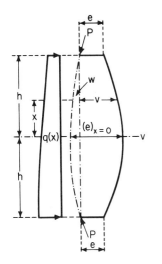

Figure H.14 □ *Eccentrically compressed column subjected to a lateral load.*

Calculation details can be found in [H.10]. The results are shown in Fig. H.13. The axial load $P = P_E$ is the Euler buckling load for the wall, and d is the wall thickness. The solid curve shows the horizontal load H divided by the axial load P times the slenderness h/d, while the dotted curve shows the horizontal load divided by the Euler buckling load P_E times the "slenderness." The first curve gives better accuracy for calculations, while the second gives a picture of the actual magnitude of the horizontal force H. The ordinate can also be thought of as a moment since $(H/2) \times (h/3) = Hh/6$ is the moment at mid height of the wall. Hh/Pd can be replaced by $6M/Pd$ and Hh/P_Ed by $6M/P_Ed$; furthermore, if $\sigma = M/W = M(6/d^2)$ and $\sigma_E = P_E/d$, the value $6M/P_E$ can be replaced by σ/σ_E; if $P/d = \sigma_0$, $P/P_E = \sigma_0/\sigma_E$. The diagram can, therefore, be thought of as a relationship between the relative stresses σ/σ_0 and σ_0/σ_E.

Figure H.13 can also be interpreted as an interaction diagram with the moment (or nominal stress from moment) on the ordinate and the axial load (or average stress) on the abscissa. It is presumed for the results given in Fig. H.13, however, that there is no risk of crushing of the material. This holds only for relatively slender walls. For less slender walls with high strength mortars, the problem has been studied and solved approximately by means of interaction diagrams (Pfrang, Grenley, and Cattaneo [H.12]). The more general and more intricate case when the axial load is eccentrically applied, the lateral load is distributed, and the risk of crushing is also taken into account, is partially dealt with below, mainly following a paper by Hellers [H.13].

H.6c: Eccentrically compressed column subjected to a lateral load: symmetrical case

Figure H.15 □ *The column part between the sections h and x.*

On the column shown in Fig. E.1, we now apply a lateral load $q(x)$ (see Fig. H.14). The pressure line through the column is then no longer straight. Let w denote the occurred displacement and let M denote the bending moment of the lateral load with respect to the section x. These functions are associated, which is clear from a moment equation for the part of the column show in Fig. H.15. Taking the moments about the axis C, it is found that

$$P \cdot w = M \tag{H.6}$$

In this case Eqs. (E.2) and (E.3) are still valid. Equation (E.1) must be modified due to the influence of w.

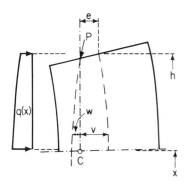

$$\eta - \frac{d}{2} + v = \frac{2}{3}\eta - w$$

which yields

$$\eta = 3\left(\frac{d}{2} - v - w\right) \tag{H.7}$$

Inserting Eqs. (E.2) and (H.7) into Eq. (E.3), a new differential equation for the elastic line of the column is obtained:

$$\frac{d^2v}{dx^2} = -\frac{2P}{9Eb} \cdot \frac{1}{\left(\dfrac{d}{2} - v - w\right)^2} \tag{H.8}$$

The differential equation (H.8) is valid wherever $\eta \leq d$.

For the special case when P is constant, we temporarily adopt the notations v_1 and v_2. The subscripts 1 and 2 refer to the absence or presence of lateral load, respectively. If $w > 0$, it is clear that $v_2 > v_1$. From a comparison of the differential equations (H.5) and (H.8) it is then evident that

$$\frac{d^2v_2}{dx^2} > \frac{d^2v_1}{dx^2}$$

Denoting the radii of curvature by ρ_1 and ρ_2, it follows that $\rho_2 < \rho_1$. Using analytic functions, the differential equation (H.8) can be solved provided that w is a polynomial of not more than second order (x^2). Concentrated lateral forces yield a first-order w, while a uniform lateral load yields a second-order w. The treatment of Eq. (H.8) in case of concentrated lateral forces is omitted here. The method applied in solving Eq. (H.8) in the case of a second-order w is applicable also to any simpler case. Let $q(x) = q = $ constant. The deflected column is then symmetrical with respect to $x = 0$. In the treatment we can therefore confine ourselves to one-half of the column.

ⓐ First loading case: $e \geq d/6$ In this case, the longitudinal forces at the column ends are applied outside or on the kern boundary. Disregarding the end sections, the pressure line in this case pierces the cross sections outside the kern boundary, implying that Eq. (H.8) is valid througout the column.

Equation (H.6) yields

$$w = \frac{qb}{2P}(h^2 - x^2)$$

Differentiating twice we obtain

$$\frac{d^2w}{dx^2} = -\frac{qb}{P}$$

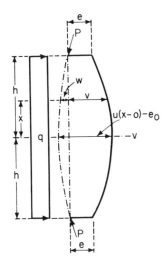

Figure H.16 □ Eccentrically
compressed column subjected to a
uniform lateral load.

The following expression can then be formed:

$$\frac{d^2 v}{dx^2} + \frac{d^2 w}{dx^2} = \frac{d^2}{dx^2}(v + w)$$

$$= -\frac{2P}{9Eb} \cdot \frac{1}{\left[\frac{d}{2} - (v + w)\right]^2} - \frac{qb}{P}$$

Substituting in this expression

$$v + w = u$$

where u is a new function denoting the distance from the pressure line to the deflected axis of the column, the following differential equation for u is obtained:

$$\frac{d^2 u}{dx^2} = -\frac{2P}{9Eb} \cdot \frac{1}{\left(\frac{d}{2} - u^2\right)} - \frac{qb}{P} \tag{H.9}$$

This equation will now be solved. It should be noticed that we need not make any restrictions concerning the direction, that is, the sign, of the lateral load q.

Multiplying by du/dx and integrating, we obtain

$$\left(\frac{du}{dx}\right)^2 = \frac{4P}{9Eb}\left(C_1 - \frac{1}{\frac{d}{2} - u}\right) - 2\frac{qb}{P} \cdot u$$

For symmetrical reasons, $du/dx = 0$ for $x = 0$. At this section, we put (Fig. H.16)

$$u(x = 0) = e_0$$

After the determination of C_1, we have

$$\left(\frac{du}{dx}\right)^2 = \frac{4P}{9Eb}(e_0 - u)\left[\frac{1}{\left(\frac{d}{2} - e_0\right)\left(\frac{d}{2} - u\right)} + \frac{9}{2}Eq\frac{b^2}{P^2}\right] \tag{H.10}$$

The two factors in Eq. (H.10)

$$(e_0 - u) \quad \text{and} \quad \left[\frac{1}{\left(\frac{d}{2} - e_0\right)\left(\frac{d}{2} - u\right)} + \frac{9}{2}Eq\frac{b^2}{P^2}\right]$$

are continuous functions of u. Since $(du/dx)^2 \geq 0$, it is evident that they are simultaneously positive, zero, or negative. For the case $q = 0$, $P > 0$, they are both positive. The two factors can both be zero only when $q = 0$ and $P = 0$. This can be seen by insertion of $e = e_0$ in Eq. (H.23), which is the solution to the case with positive factors. It is therefore possible to transit to the negative case only via $P = 0$. From this it is obvious that a longitudinally loaded column cannot remain in equilibrium, in a negative case, if the lateral load is removed. Such a case has no practical significance and is therefore excluded from further treatment of the theory.

The first derivative obtained from Eq. (H.10) can hence be written as follows:

$$\frac{du}{dx} = \mp \frac{2}{3}\sqrt{\frac{P}{Eb}}\frac{1}{\sqrt{\frac{d}{2}-e_o}}\sqrt{\frac{e_0-u}{\frac{d}{2}-u}} \tag{H.11}$$

$$\times \sqrt{1+\frac{9}{2}Eq\frac{b^2}{P^2}\left(\frac{d}{2}-e_0\right)\left(\frac{d}{2}-u\right)}$$

The variables in Eq. (H.11) can be separated. After a formal integration, and considering the boundary conditions, we obtain

$$x = \pm\frac{3}{2}\sqrt{\frac{Eb}{P}}\sqrt{\frac{d}{2}-e_0}\int_{u(x)}^{e_0}\sqrt{\frac{\frac{d}{2}-u}{e_0-u}}$$

$$\times \frac{du}{\sqrt{1+\frac{9}{2}Eq\frac{b^2}{P^2}\left(\frac{d}{2}-e_0\right)\left(\frac{d}{2}-u\right)}} \tag{H.12}$$

By introduction of the substitution

$$s = \sqrt{\frac{\frac{d}{2}-u}{e_0-u}} \qquad s \geq 1 \tag{H.13}$$

and the convenient parameters

$$e_0 = \frac{d}{6}+p\frac{d}{3} \tag{H.14}$$

$$m = 6\frac{e}{d} \tag{H.15}$$

$$R = \frac{1}{4}Eq\left(\frac{bd}{P}\right)^2 \tag{H.16}$$

Eq. (H.12) reads

$$x = \pm 2\sqrt{\frac{EI}{P}}(1-p)^{3/2}$$

$$\times \int_{s(u(x))}^{\infty}\frac{s^2\,ds}{(s^2-1)^{3/2}\sqrt{[1+2R(1-p)^2]s^2-1}} \tag{H.17}$$

where I denotes the flexural rigidity of the column

$$I = \frac{bd^3}{12}$$

With the further substitution

$$s = \frac{1}{\sin t} \qquad (H.18)$$

the equation is transferred into the suitable form

$$x = \pm 2 \sqrt{\frac{EI}{P}} (1 - p)^{3/2} k \int_0^{t\{s[u(x)]\}} \frac{dt}{\cos^2 t \sqrt{1 - k^2 \sin^2 t}} \qquad (H.19)$$

where

$$k = \frac{1}{\sqrt{1 + 2R(1 - p)^2}} \qquad (H.20)$$

The integral in Eq. (H.19) is elliptical and may be expressed using elliptic basic integrals. Let us abbreviate by setting

$$t\{s[u(x)]\} = t(x)$$

The integral in Eq. (H.19) can then be expressed as

$$\int_0^{t(x)} \frac{dt}{\cos^2 t \sqrt{1 - k^2 \sin^2 t}}$$

$$= \frac{1}{1 - k^2} \{\Psi[k, t(x)] - E[k, t(x)] + (1 - k^2)F[k, t(x)]\}$$

where

$$\left. \begin{array}{l} \Psi[k, t(x)] = \tan t(x) \cdot \sqrt{1 - k^2 \sin^2 t(x)} \\[2mm] F[k, t(x)] = \int_0^{t(k)} \frac{dt}{\sqrt{1 - k^2 \sin^2 t}} \quad \left(\begin{array}{l} \text{incomplete elliptic} \\ \text{integral of the first} \\ \text{order} \end{array}\right) \\[4mm] E[k, t(x)] = \int_0^{t(x)} \sqrt{1 - k^2 \sin^2 t}\, dt \quad \left(\begin{array}{l} \text{incomplete elliptic} \\ \text{integral of the sec-} \\ \text{ond order} \end{array}\right) \end{array} \right\} \qquad (H.21)$$

For $R > 0$, an angle α may be defined

$$\alpha = \text{arc sin } k$$

This angle is important in calculations depending on mathematical tables since incomplete elliptic integrals are normally tabulated by $[\alpha, t(x)]$. In case $R < 0$, it follows that $k > 1$, see Eq. (H.20). Using mathematical tables, we must then resort to the transformations

$$F[k, t(x)] = \int_0^{t(x)} \frac{dt}{\sqrt{1 - k^2 \sin^2 t}}$$

$$= k^{-1} \cdot F[k^{-1}, \theta(x)]$$

$$E[k, t(x)] = \int_0^{t(x)} \sqrt{1 - k^2 \sin^2 t}\, dt$$

$$= k \cdot E[k^{-1}, \theta(x)] - (k^2 - 1) \cdot F[k, t(x)]$$

where

$$\theta(x) = \text{arc sin } [k \sin t(x)]$$

to be able to introduce the angle α, now defined as

$$\alpha = \text{arc sin } (k^{-1})$$

We have now obtained the following solution of the differential equation (H.9):

$$x = \pm 2 \sqrt{\frac{EI}{P}} (1 - p)^{3/2} \frac{k}{1 - k^2} \{\Psi[k, t(x)] - E[k, t(x)]$$

$$+ (1 - k^2)F[k, t(x)]\} \tag{H.22}$$

where $t(x)$ may be derived from the substitutions in Eqs. (H.13) and (H.18) and k is taken according to Eq. (H.20). The function symbols Ψ, F, and E are defined in the expressions in Eq. (H.21). Equation (H.22) is positive for $x > 0$ and negative for $x < 0$. The upper limit of the angle $t(x)$ is denoted by φ. The limit is reached for $x = \pm h$. Inserting $x = h$ into Eq. (H.22), we obtain the important formula

$$h\sqrt{\frac{P}{EI}} = 2(1 - p)^{3/2} \frac{k}{1 - k^2}$$

$$\cdot \{\Psi[k, \varphi] - E[k, \varphi] + (1 - k^2)F[k, \varphi]\}$$

Using the parameters according to Eqs. (H.14) and (H.15), we find that

$$\varphi = \text{arc sin } \sqrt{\frac{1 + 2p - m}{3 - m}} \tag{H.23}$$

Noting that k, Eq. (H.20), is a function of p and R, it can now be concluded that the formula above defines a force function

$$h\sqrt{\frac{P}{EI}} = f_1(p, m, R)$$

It can be shown that Eq. (H.23) approaches, as $q \to 0$, the corresponding formula for $h\sqrt{P/EI}$ for $q = 0$ given by Angervo in [E.1] (see Chapter E). In the further treatment a formula for the derivative at the column ends will prove useful. By insertions in Eq. (H.11) we find

$$\left(\frac{du}{dx}\right)_{u=e} = \mp \frac{d}{3} \sqrt{\frac{P}{EI}} \sqrt{\frac{1 + 2p - m}{(1 - p)(3 - m)}} \tag{H.24}$$

$$\times \sqrt{1 + R(1 - p)(3 - m)}$$

The formulas in Eqs. (H.23) and (H.24) are valid when

$$1 \leq m \leq 3 \atop \dfrac{m-1}{2} \leq p \leq 1 \Bigg\} \qquad \text{(H.25)}$$

Furthermore, the functions in the formula in Eq. (H.23) must have real values which lead to the restriction

$$k^2 \sin^2 \varphi \leq 1$$

where k and φ are defined according to Eqs. (H.20) and (H.23), respectively. This condition can be written in the form

$$R \geq -\frac{1}{(1-p)(3-m)} \qquad \text{(H.26)}$$

When R passes its lower limit, the column breaks through implying a total failure. At the limit, Eq. (H.24) shows that

$$\left(\frac{du}{dx}\right)_{u=e} = 0$$

ⓑ Second loading case : $e < d/6$ at the end parts of the column only   In this case, the pressure line pierces the cross sections outside the kern boundary in the center part of the column only. Let $x = x^*$ denote the section on the upper half of the column where $u = d/6$. The height h is divided into two parts by this section

$0 \leq x \leq x^*$,   where $\eta \leq d$; i.e., the cross section is partly ruptured;

$x^* \leq x \leq h$,   where the whole cross section is active.

For the part $0 \leq x \leq x^*$, the differential equation (H.9) is valid. The slope of the pressure line toward the column axis for $x = x^*$ is found by inserting $u = d/6$ into Eq. (H.11).

$$\left(\frac{du}{dx}\right)_{x=x^*} = -\frac{d}{3}\sqrt{\frac{P}{EI}}\sqrt{\frac{p}{1-p}}\sqrt{1+2R(1-p)} \qquad \text{(H.27)}$$

For the part $x^* \leq x \leq h$, we have the well-known differential equation

$$\frac{d^2v}{dx^2} = -\frac{P(v+w)}{EI} \qquad \text{(H.28)}$$

By substituting

$$v + w = u$$

we obtain

$$\frac{d^2u}{dx^2} = -\frac{Pu}{EI} - \frac{qb}{P} \qquad \text{(H.29)}$$

Performing the quadrature as in the previous case ⓐ, the first derivative is obtained:

$$\frac{du}{dx} = \mp \sqrt{-\frac{P}{EI}u^2 - \frac{qb}{P}2u + C_2}$$

(H.30)

By inserting $u = d/6$ into Eq. (H.30) and requiring that the obtained expression equals Eq. (H.27), C_2 is determined. Inserting the expression for C_2 and transforming, we obtain

$$\frac{du}{dx} = \mp \sqrt{\frac{P}{EI}} \sqrt{\frac{d^2}{36}\left(\frac{1+3p}{1-p} + 8pR + 4R\right) - \frac{2}{3}R\,du - u^2}$$

(H.31)

The continued integration must be performed with respect to the two parts of the half column, separated by the section x^*. For the part $0 \leq x \leq x^*$, the treatment can be based on what was shown in case ⓐ.

The following limits apply:

$$s(x=0) = \infty, \qquad s(x=x^*) = \frac{1}{\sqrt{p}}$$

where s is the variable defined in Eq. (H.13). Therefore, according to Eq. (H.22),

$$x^* = 2\sqrt{\frac{EI}{P}}(1-p)^{3/2}\frac{k}{1-k^2}$$

$$\times \{\Psi[k,\varphi] - E[k,\varphi] + (1-k^2)F[k,\varphi]\}$$

(H.32)

where

$$\varphi = \text{arc sin }\sqrt{p}$$

For the part $x^* < x < h$, we return to Eq. (H.31). After a separation of variables and a formal integration, the equation can be written in the form (where $x = h$ has been inserted)

$$\sqrt{\frac{EI}{P}}\int_{md/6}^{d/6}\frac{du}{\sqrt{\frac{d^2}{36}\left[\frac{1+3p}{1-p} + 4R(R+2p+1)\right] - \left(u + \frac{Rd}{3}\right)^2}}$$

$$= \int_{x^*}^{h} dx$$

This yields

$$\sqrt{\frac{EI}{P}}\,I_{md/6}^{d/6}\text{ arc sin }\frac{u + \frac{Rd}{3}}{\frac{d}{6}\sqrt{\frac{1+3p}{1-p} + 4R(R+2p+1)}} = I_{x^*}^{h}x$$

Taking x from Eq. (H.32), we finally obtain the important formula

$$h\sqrt{\frac{P}{EI}} = 2(1 - p)^{3/2} \frac{k}{1 - k^2}$$

$$\times \{\Psi[k, \varphi] - E[k, \varphi] + (1 - k^2)F[k, \varphi]\}$$

$$+ \arcsin \frac{1 + 2R}{\sqrt{\dfrac{1 + 3p}{1 - p} + 4R(R + 2p + 1)}}$$

$$- \arcsin \frac{m + 2R}{\sqrt{\dfrac{1 + 3p}{1 - p} + 4R(R + 2p + 1)}} \qquad \text{(H.33)}$$

where

$$\varphi = \arcsin \sqrt{p} \qquad \text{(H.33a)}$$

This formula defines a force function

$$h\sqrt{\frac{P}{EI}} = f_2(p, m, R)$$

In further treatment, a formula for the derivative equation (H.31) at the column ends will prove useful. By inserting, we find

$$\left(\frac{du}{dx}\right)_{u=e} = \mp \frac{d}{6}\sqrt{\frac{P}{EI}}\sqrt{\frac{1 + 3p}{1 - p} + 4R(1 + 2p - m) - m^2}$$

$$\text{(H.34)}$$

The formulas in Eqs. (H.33) and (H.34) are valid when

$$\left.\begin{array}{l} 0 \leq m < 1 \\ 0 \leq p \leq 1 \end{array}\right\} \qquad \text{(H.35)}$$

As in case ⓐ, we must consider an additional condition of stability. The real-valued arc sine functions in Eq. (H.33) are defined when

$$R \geq -\frac{1}{4} \cdot \frac{\dfrac{1 + 3p}{1 - p} - m^2}{1 + 2p - m} \qquad \text{(H.36)}$$

This condition is more rigorous than the similar condition of the elliptic integrals in Eq. (H.33). Inserting Eq. (H.36) into Eq. (H.34), we see that equality in Eq. (H.36) implies that

$$\left(\frac{du}{dx}\right)_{u=e} = 0$$

© Third loading case: $e < d/6$ In this case the pressure line of the forces pierces all cross sections within the kern boundary. Here Eq. (H.29),

$$\frac{d^2u}{dx^2} = -\frac{P \cdot u}{EI} - \frac{qb}{P}$$

is valid throughout the column. Integration yields

$$\left(\frac{du}{dx}\right)^2 = -\frac{P}{EI} \cdot u^2 - \frac{qb}{P} 2u + C_3 \tag{H.37}$$

With aid of the boundary condition

$$\left(\frac{du}{dx}\right)_{u=e_0} = 0$$

C_3 is determined. We obtain

$$\left(\frac{du}{dx}\right)^2 = \frac{P}{EI}(e_\bullet - u)\left(e_0 + u + \frac{2}{3}Rd\right) \tag{H.38}$$

After a discussion of the two factors in Eq. (H.38), $(e_0 - u)$ and $(e_0 + u + \frac{2}{3}Rd)$, analogous to the corresponding discussion in case ⓐ, we conclude that the positive combination implying that $u \leq e_0$ must be chosen and consequently

$$\frac{du}{dx} = \mp\sqrt{\frac{P}{EI}}\sqrt{e_0 - u}\sqrt{e_0 + u + \frac{2}{3}Rd} \tag{H.39}$$

After a separation of variables and a formal integration (inserting $x = h$), we obtain

$$\sqrt{\frac{EI}{P}}\int_0^{e_0}\frac{du}{\sqrt{e_0 - u}\sqrt{e_0 + u + \frac{2}{3}Rd}} = \int_0^h dx \tag{H.40}$$

and if we introduce the substitution $\sqrt{e_0 - u} = y$ equation (H.40) is transformed into

$$\sqrt{\frac{EI}{P}}\int_0^{\sqrt{e_0-e}}\frac{2\,dy}{\sqrt{2e_0 + \frac{2}{3}Rd - y^2}} = \int_0^h dx \tag{H.41}$$

Performing the integration and inserting the parameters according to Eqs. (H.14) and (H.15), we obtain the important formula

$$h\sqrt{\frac{P}{EI}} = 2 \text{ arc sin}\sqrt{\frac{1 + 2p - m}{2(1 + 2p + 2R)}}$$

For brevity, the notation of a force function

$$h\sqrt{\frac{P}{EI}} = f_3(p, m, R) \tag{H.42}$$

is introduced.

As in the previous cases ⓐ and ⓑ, a formula for the derivative equation (H.39) at the column ends proves useful in the further treatment:

$$\left(\frac{du}{dx}\right)_{u=e} = \mp \frac{d}{6}\sqrt{\frac{P}{EI}}\sqrt{(1+2p+m)(1+2p+m+4R)}$$

(H.43)

The formulas in Eqs. (H.42) and (H.43) are valid when

$$\left.\begin{array}{c} 0 \leq m < 1 \\ \dfrac{m-1}{2} \leq p \leq 0 \end{array}\right\}$$

(H.44)

As in the case ⓐ and ⓑ, we must here employ an additive condition of stability. The real-valued arc sine functions in Eqs. (H.42) are defined when

$$R \geq \tfrac{1}{4} \cdot (1+2p+m)$$

(H.45)

The case of equality in Eq. (H.45) implies that

$$\left(\frac{du}{dx}\right)_{u=e} = 0$$

which is clear from inserting the limit into Eq. (H.43).

With these three symmetrical loading cases ⓐ, ⓑ, and ⓒ as a basis, it is also possible to treat case of column under asymmetric load.

H.6d : Eccentrically compressed column subjected to a uniform lateral load : unsymmetrical case. Design diagrams

All the formulas necessary for a mathematical study of the behavior of a laterally loaded wall with eccentric axial load were derived in Section H.6c. To obtain a solution for an unsymmetrical case, a combination of solutions is worked out in basically the same way as demonstrated in Fig. E.2, although the derivations are more elaborate because of the new variable q. The details can all be found in Hellers' paper [H.13]. In this chapter some of the practical results are pointed out.

First the behavior of a wall having no tensile stress, eccentrically loaded at one end, concentrically loaded at the other end, and loaded with a uniform lateral load, will be described. As shown in Fig. E.8, for a wall of this type the relationship between the applied load and the end rotation follows to the dotted line in

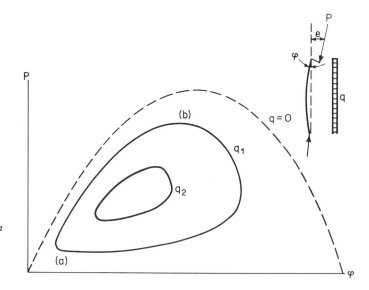

Figure H.17 ☐ Relationship between angle of rotation of a wall end (φ) and the eccentric axial load (P) for varying lateral load (q).

Figure H.18 ☐ Buckling diagram for eccentrically loaded wall without lateral load (q = 0). d = wall thickness.

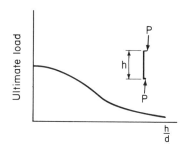

Fig. H.17 if the lateral load q is zero. When $q > 0$ the axial load (solid curves) must be positive definite before a position of equilibrium is possible [the minimum point on the solid curves (*a*)]. As in the case with $q = 0$, there is also a maximum axial load for which a solution exists [point (*b*) on the curve]. As the load increases from (*a*) to (*b*), the stresses also increase, and it is quite possible that the wall will fail in compression. This case is discussed later. It can be seen from Fig. H.17 that the wall has for each q, as in the case with $q = 0$, two possible equilibrium positions, one stable and one unstable with the same magnitudes of P.

As the lateral load increases, the region of P in which the wall is stable decreases, and the closed curve describing the relationship between the axial load and the end rotation finally becomes a point for a lateral load $q = q_{\text{ult}}$ which cannot be exceeded. This lateral load q_{ult} can be reached only for one specific axial load for given dimensions and elastic constants of the wall.

For design purposes it is convenient to display the results from the calculations (in which the compressive strength of the material is also taken into account) in the form of buckling diagrams of the well known type shown in Fig. H.18. In the case when a lateral load is applied, the diagram shown in Fig. H.18 is modified since the h/d ratio can no longer be increased in-

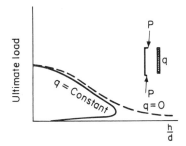

Figure H.19 □ Buckling diagram for an eccentrically loaded wall with lateral load q.

finitely, as it can without a lateral load. The diagram now takes the shape shown in Fig. H.19, in which the diagram from Fig. H.18 is shown for comparison. Buckling diagrams of the type discussed have been calculated by Hellers [H.13] for a set of end eccentricities as shown in Figs. H.20 to H.25. In these diagrams, σ_B is the strength of the wall material in compression, ϵ_B is the strain at σ_B. This strain is $\epsilon_B = \sigma_B/E$ since the material is assumed to be linearly elastic in compression.

Figure H.20 □ Buckling diagram for an eccentrically loaded wall with lateral load q. For an interpretation of the diagram, see Fig. H.19. [H.13].

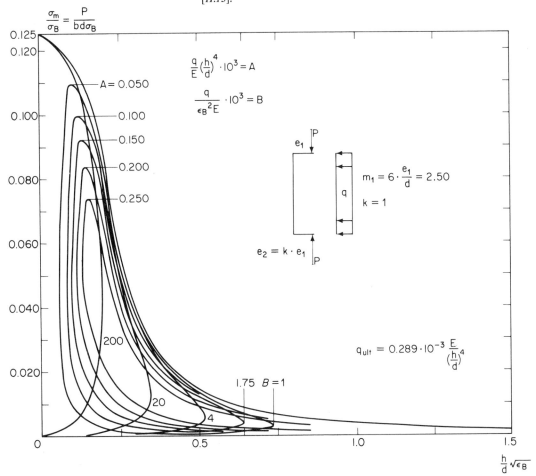

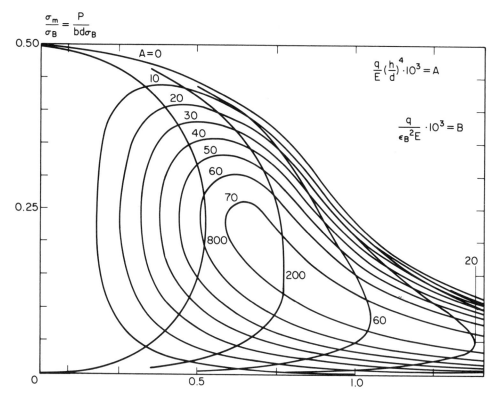

$$\frac{\sigma_m}{\sigma_B} = \frac{P}{bd\sigma_B}$$

A = 0

$$\frac{q}{E}\left(\frac{h}{d}\right)^4 \cdot 10^3 = A$$

$$\frac{q}{\epsilon_B^2 E} \cdot 10^3 = B$$

Figure H.21 □ *Buckling diagram for an eccentrically loaded wall with lateral load q. For an interpretation of the diagram, see Fig. H.19.* [*H.13*].

Figure H.22 □ *Buckling diagram for an eccentrically loaded wall with lateral load q. For an interpretation of the diagram, see Fig. H.19.* [*H.13*].

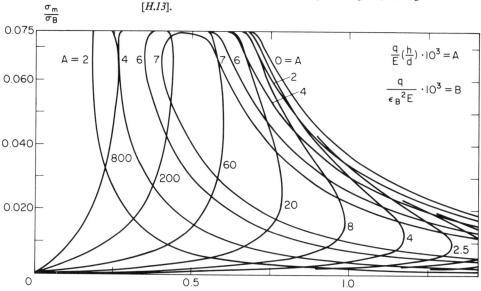

$$\frac{\sigma_m}{\sigma_B}$$

$$\frac{q}{E}\left(\frac{h}{d}\right) \cdot 10^3 = A$$

$$\frac{q}{\epsilon_B^2 E} \cdot 10^3 = B$$

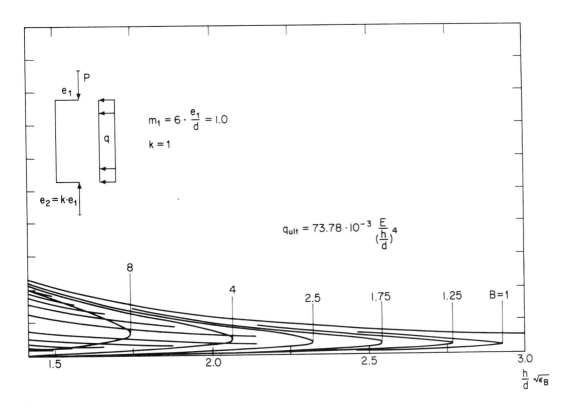

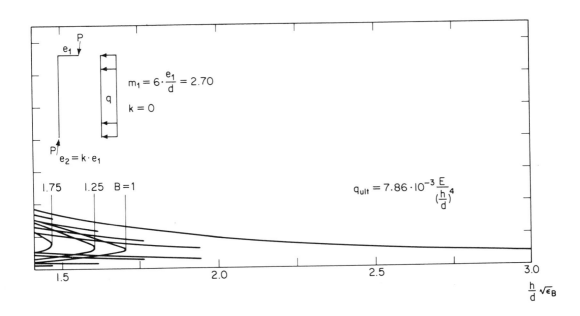

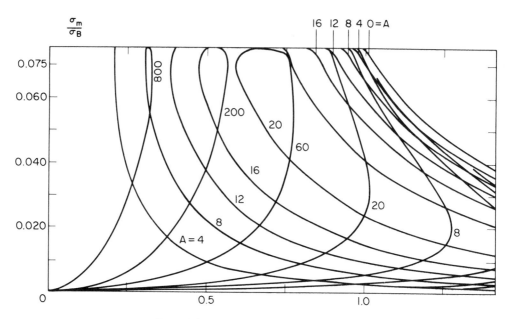

Figure H.23 □ Buckling diagram for an eccentrically loaded wall with lateral load q. For an interpretation of the diagram, see Fig. H.19. [H.13].

Figure H.24 □ Buckling diagram for an eccentrically loaded wall with lateral load q. For an interpretation of the diagram, see Fig. H.19. [H.13].

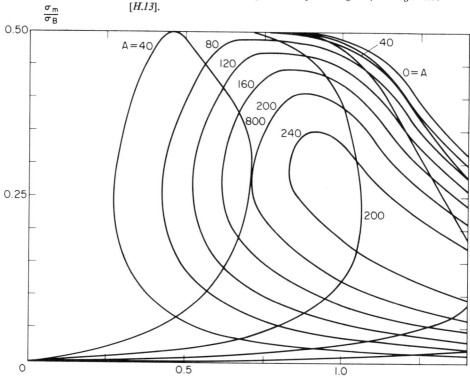

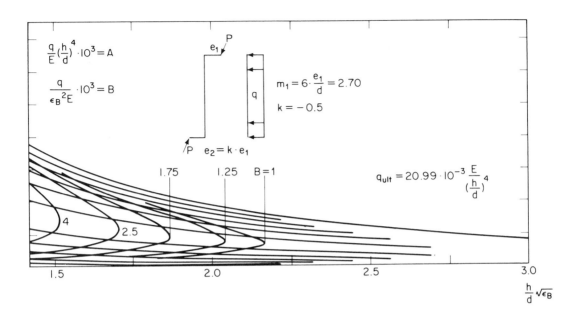

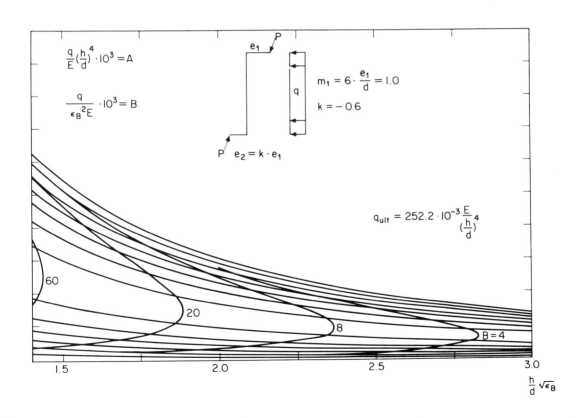

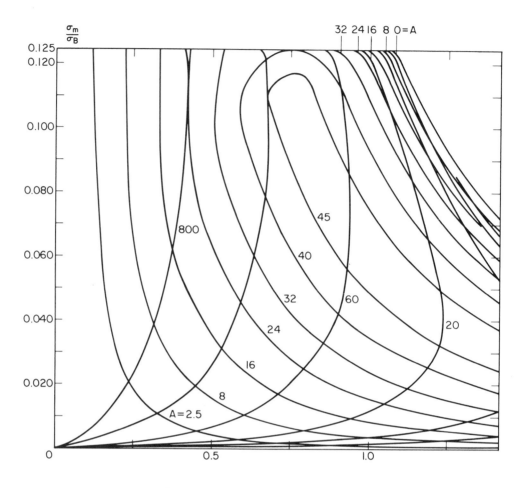

Figure H.25 ☐ *Buckling diagram for an eccentrically loaded wall with lateral load q. For an interpretation of the diagram, see Fig. H.19.* [*H.13*].

H.7 ☐ Interaction of an eccentrically loaded wall subjected to lateral loads with slabs

H.7a : Introductory remarks

In Sections H.6c and d the problem of a statically determinant case of wall loading has been treated. This section deals with the problem when the wall interacts with slabs at the ends and thus creates a statically indeterminate case in which the end eccentricities are unknown a priori and have to be calculated. This can at least theoretically be done by studying the continuity at the joints. The angles of rotation must therefore be studied. Again the paper [H.13] by Hellers is followed.

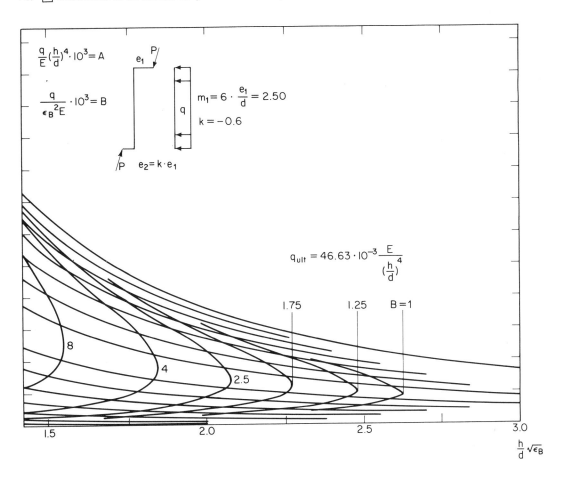

$$\frac{q}{E}\left(\frac{h}{d}\right)^4 \cdot 10^3 = A$$

$$\frac{q}{\epsilon_B^2 E} \cdot 10^3 = B$$

$$m_1 = 6 \cdot \frac{e_1}{d} = 2.50$$

$$k = -0.6$$

$$e_2 = k \cdot e_1$$

$$q_{ult} = 46.63 \cdot 10^{-3} \frac{E}{\left(\frac{h}{d}\right)^4}$$

H.7b: The rotations occurring at the ends of a column leg of a frame

Consider a column between two stories of a building acting as a leg of a frame as shown in Figs. H.26 and H.27. Figure H.26 belongs to the case where the curvature of the elastic line maintains its sign throughout the column, whereas in the case of Fig. H.27 the curvatures at the column ends have opposite signs. In addition to the previous notations, c_1 and c_2 have been used to denote the displacements of the joints, in case the system is displaceable. Furthermore, we have introduced the angle α_1 denoting the slope at the ends of the pressure line of the exterior forces with regard to the line of action of the longitudinal forces only. We find for α_1, using Eq. (H.16), the formula

$$\frac{h}{d} \tan \alpha_1 = \frac{qbh^2}{2Pd} = \frac{R}{6} \left(h\sqrt{\frac{P}{EI}} \right)^2$$

A consideration based on Fig. H.26 reveals that the rotation of joint 1 can be written

$$\varphi_{v_1} = \text{arc tan} \frac{e_1 - e_2}{h} - \text{arc tan} \frac{c_1 - c_2}{h}$$

$$- \alpha_1 \pm \text{arc tan} \left| \frac{du}{dx} \right|_{u=e_1} \qquad (H.46)$$

whereas in the case shown in Fig. H.27,

$$\varphi_{v_1} = \text{arc tan} \frac{e_1 + e_2}{h} - \text{arc tan} \frac{c_1 - c_2}{h} - \alpha_1 \pm \text{arc tan} \left| \frac{du}{dx} \right|_{u=e_1}$$

$$(H.47)$$

In these formulas the sign of the last member is $+$ or $-$, according to whether the curvature of the elastic line increases or decreases from the joint 1 downward.

Equations (H.46) and (H.47) can be brought into a more suitable form by inserting the relative eccentricities m_1 and m_2 and by introducing the notation

$$\psi = \text{arc tan} \frac{c_1 - c_2}{h}$$

for the rotation of the column axis caused by the displacement of the joints. The angles shown in Eqs. (H.46) and (H.47) can be replaced by their tangents without any considerable error. If this is done, one obtains the useful formula

$$\frac{h}{d}(\varphi_{v_1} + \psi) = \frac{m_1 \pm m_2}{6} - \frac{R}{6} \left(h\sqrt{\frac{P}{EI}} \right)^2 \pm \frac{h}{d} \left| \frac{du}{dx} \right|_{u=e_1}$$

$$(H.48)$$

One of the three expression in Eqs. (H.24), (H.34), or (H.43) for (du/dx) at the column end has to be used in Eq. (H.48).

5.1	5.1.1	5.1.2	5.1.3	5.1.3.1	5.1.3.2	5.2	5.2.1	5.2.2
The eccentricity has the same sign at both ends of the column	$m_1 > 1$ $m_2 \geq 1$ $m_1 > m_2$	$m_1 > 1$ $m_2 < 1$	$m_1 \leq 1$ $m_2 < 1$ $m_1 > m_2$	The pressure line inside the kern	The pressure line outside the kern in some parts	*The eccentricity has opposite sign at column ends*	$m_1 \geq 1$ $m_2 \geq 1$	$m_1 \geq 1$ $m_2 < 1$

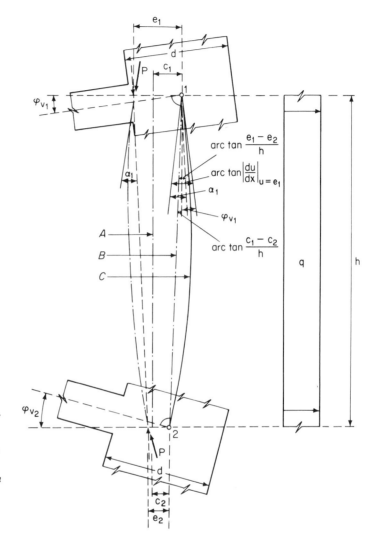

Figure H.26 □ A vertical member of a frame, its curvature having the same sign at both ends of the leg. A. Position of the column axis before displacement; B. after rigid body motion; C. after bending. (Note that the line of action near joint 1 is incorrectly drawn outside the column thickness to avoid interfering with the figure.)

Table H.1 □ Validity conditions for equations (H.49), (H.50), and (H.51), from [H.13].

5.2.2.1	5.2.2.2	5.2.3	5.2.3.1	5.2.3.2	5.2.3.2.1	5.2.3.2.2	5.2.3.2.3
The pressure line inside the kern in the lower part of the column	The pressure line partly outside the kern in the lower part of the column	$m_1 < 1$ $m_2 < 1$	The pressure line inside the kern in the whole column	The pressure line partly pierces cross sections outside the kern	In the upper part inside; in the lower parts partly outside the kern	In the upper part partly outside; in the lower parts inside the kern	In both parts partly outside the kern

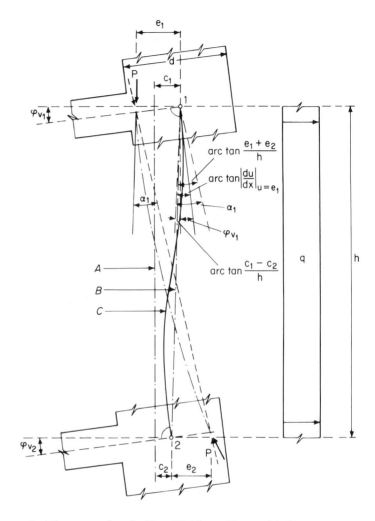

Figure H.27 □ A vertical member of a frame, its curvature having opposite signs at the ends. A. Position of the column axis before displacement; B. after rigid body motion; C. after bending.

1. The expression in Eq. (H.24) applies to $(du/dx)_{u=e_1}$. We obtain, by inserting Eq. (H.24) into Eq. (H.48),

$$\frac{h}{d}(\varphi_{v_1} + \psi) = \frac{m_1 \pm m_2}{6} - \frac{R}{6}\left(h\sqrt{\frac{P}{EI}}\right)^2$$
$$\pm \frac{1}{3}h\sqrt{\frac{P}{EI}}\sqrt{\frac{1 + 2p_1 - m_1}{(1 - p_1)(3 - m_1)}}$$
$$\times \sqrt{1 + R(1 - p_1)(3 - m_1)} \qquad \text{(H.49)}$$

This formula is valid in the cases ⓐ: 5.1.1, 5.1.2, 5.2.1, and 5.2.2 as defined in Table H.1. See also Section H.6c. The numbers refer to Hellers' original numbering of the cases.

2. The expression in Eq. (H.34) applies to $(du/dx)_{u=e_1}$. We obtain, by inserting Eq. (H.34) into Eq. (H.48),

$$\frac{h}{d}(\varphi_{v_1} + \psi) = \frac{m_1 \pm m_2}{6} - \frac{R}{6}\left(h\sqrt{\frac{P}{EI}}\right)^2$$
$$\pm \frac{1}{6}h\sqrt{\frac{P}{EI}}\sqrt{\frac{1 + 3p_1}{1 - p_1} + 4R(1 + 2p_1 - m_1) - m_1^2} \quad \text{(H.50)}$$

This formula is valid in the cases ⓑ: 5.1.3.2, 5.2.3.2.2, and 5.2.3.2.3.

3. The expression in Eq. (H.43) applies to $(du/dx)_{u=e_1}$. We obtain, by inserting Eq. (H.43) into Eq. (H.48),

$$\frac{h}{d}(\varphi_{v_1} + \psi) = \frac{m_1 \pm m_2}{6} - \frac{R}{6}\left(h\sqrt{\frac{P}{EI}}\right)^2$$
$$\pm \frac{1}{6}h\sqrt{\frac{P}{EI}}\sqrt{(1 + 2p_1 - m_1)(1 + 2p_1 + m_1 + 4R)}$$

$$\text{(H.51)}$$

This formula is valid in the cases ⓒ: 5.1.3.1, 5.2.3.1, and 5.2.3.2.1. The angle of rotation function $(h/d)(\varphi_{v_1} + \psi)$ is defined in the same regions as the corresponding force function in the different cases.

H.7c: The interaction problem

The problem of the column interacting with horizontal members must be solved numerically with the aid of a computer to avoid extremely· laborious graphical methods. In principle we need to know the relation of the angular end rotations to the lateral load for constant longitudinal load and variable end eccentricities. As an example, the relation is shown in a special case with constant end eccentricities in Fig. H.28. The relation has been restricted to the stable part of the solution in the statically determinate case. In the statically indeterminate case, the other part of the solution should also be considered.

H.7d: Walls of material having tensile strength

In cases where the masonry wall has an appreciable tensile strength, the problem of calculation of force distribution in a building frame consisting of walls and slabs is reduced to the common type which (at least for some basic plane structures) is treated in most textbooks on strength and stability. See, for example, Timoshenko and Young [H.14]. The application to steel columns

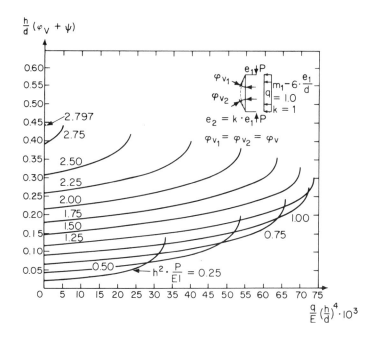

Figure H.28 □ The relation of the angular end rotations to the lateral load for constant longitudinal load, stable part of the statically determinate solution [H.13].

was treated by Young [H.15] and a more comprehensive treatment can be found in American Institute of Steel Construction [H.16]. The calculation of restrained reinforced concrete columns has been studied by Broms and Viest [H.17]. Calculations for unreinforced wall panels can be found in [H.18] and [H.19]. Since walls of material having tensile strength have been treated in several respects and at length, these solutions can readily be applied to masonry walls having tensile strength, provided the properties of the actual material are taken into account.

H.8 □ Arching action

If the wall has immovable supports, the situation is different from that described in Section H.6 and the wall can, because of the vault or arching effect, carry horizontal loads which greatly exceed those indicated in Fig. H.13 (which were calculated for a known normal force but with movement possible in the wall plane). For the more favorable case, with immovable supports, certain basic calculation factors are to be found in McDowell, McKee, and Levin [H.20], as well as in Cohen and Laing [H.21]. In the paper by McDowell, McKee, and Levin, the maximum moment has been calculated as a function of the slenderness

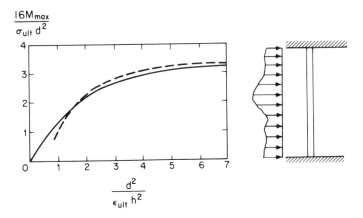

Figure H.29 □ Relationship of ultimate moment M_{max} and slenderness ratio h/d of a wall with rigid supports permitting arching action [H.11].

ratio and the ultimate strain in the wall. See the solid line in Fig. H.29. A similar calculation has not been made by Cohen and Laing. Instead, the maximum moment is given in the following equation, based on the geometrical situation and the equilibrium conditions that exist when the wall snaps through in two pieces.

$$M_{\max} = \frac{\sigma_{\text{ult}}}{4}\left(d - \frac{d\sigma_{\text{ult}}}{E\epsilon_m}\right)^2 \tag{H.52}$$

where

$$\epsilon_m = \frac{\sqrt{d^2 + \dfrac{h^2}{4}} - \dfrac{h}{2}}{\sqrt{d^2 + \dfrac{h^2}{4}}} \tag{H.53}$$

The writer has rewritten these equations with some approximations as follows:

$$\frac{4M_{\max}}{\sigma_{\text{ult}}\,d^2} \approx \left(1 - \frac{\dfrac{\sigma_{\text{ult}}}{E}}{\dfrac{h^2}{2d^2} + 1}\right)^2 \approx \left(1 - \frac{\epsilon_{\text{ult}}h^2}{2d^2}\right)^2 \quad \text{if} \quad \frac{h^2}{2d^2} \gg 1$$

$$\frac{16M_{\max}}{\sigma_{\text{ult}}\,d^2} \approx 4\left(1 - \frac{\epsilon_{\text{ult}}h^2}{2d^2}\right)^2 \quad \text{for} \quad \frac{h}{d} < 24$$

$$\tag{H.54}$$

The relationship according to Eq. (H.54) is shown with a dashed line in Fig. H.29.

The cases where bending effect occurs either in the vertical or in the horizontal plane are treated by Cohen and Laing alike, provided that the compressive strength and the ultimate strain of the material in the actual direction are inserted into the equations. In cases where the wall has a four-sided support, the load-carrying capacities originating from beam effect and vault effect in the horizontal direction are added. A more exact treatment of the problem, however, necessitates the calculation of the complete moment distribution and axial force values. Furthermore, the wall strength has to be evaluated in all directions. Attacks on the problem along these lines have been made by Pettersson [H.22], Nilsson [H.23], and Hallquist [H.6].

H.9 ☐ Slab action. Yield line theory

In cases where the axial load in the wall is small, the masonry has a certain tensile strength, and the wall is supported along at least three sides, the wall can be treated approximately as a slab. For theories on slabs the reader is referred to common textbooks (Timoshenko and Woinowsky-Krieger [H.24]). The theory of plates has been applied to masonry walls by Pettersson [H.22], who calculated diagrams for permissible dimensions of brick masonry walls with openings laterally loaded by the Swedish code wind load, and with the assumption of maximum permissible stress.

Bradshaw and Entwisle [H.25] have proposed an approximate method for calculation of the load-carrying capacity of masonry walls subjected to lateral forces only. Nilsson [H.23] has attacked the problem of lateral loads on masonry walls as a plate problem, cautiously following the yield line approach. Taking into account the orthotrophy of masonry (different moduli of elasticity in two directions), Hallquist [H.6] applied a finite element method for the calculation of the moment distribution and the stresses in a masonry wall. So far, however, his calculation only shows the x and y stresses and not the eventually inclined principal stresses. The yield line theory has been applied to laterally loaded walls by Isaacs [H.26], but again difficulties arise as to whether the deformations are compatible at maximum moment in different directions.

All the methods mentioned above can be used to estimate the plate or slab action of a laterally loaded masonry wall, but they should be used with some caution since the basic problem of the behavior of a masonry wall in biaxial bending has been studied relatively little; therefore, the strength, stiffness, and ductility in different directions can be only roughly estimated.

Table H.2 ☐ *Compilation of results from calculated examples of a 12.5 cm (5 in.) wall with a lateral load q [H.9]. All calculations were made with the following assumed wall characteristics:*

Height h = 250 cm = 8 ft 4 in.
Length l = 400 cm = 13 ft 3 in.
Thickness d = 12.5 cm = 5 in.
Modulus of elasticity
  E = 7000 kg/cm² = 100,000 psi
Ultimate strain of wall material
  $\epsilon_B = 0.004 = 4$ per mil
Vertical (axial) load in the wall
  P = 2 tons/m = 1340 lb/ft

Example number	q_{ult} kg/m²	q_{ult} lb/ft²	Span	Assumptions	Method of solution
1	85	17	h	$\sigma_{tens} = 0$	Figure H.13 with distributed load q.
2	166	34	h	$\sigma_{tens} = 4\,kg/cm^2$	Theory of elasticity for a beam with axial load and lateral load q.
3	75	15	h	$\sigma_{tens} = 0$ Eccentric axial load $e_1 = 2.1$ cm and $e_2 = 1.25$ cm	Approximative method based on Figures E.12 and optimum eccentricities at failure.
4	154	31	h	$\sigma_{tens} = 4\,kg/cm^2$ Eccentric axial load $e_1 = 2.1$ cm and $e_2 = 1.25$ cm	Theory of elasticity for a beam with axial load and lateral load q.
5	570	116	h	Arching action one way	Figure H.29.
6	622	127	h, l	Arching action two ways	Figure H.29.
7	200	41	h	$\sigma_{tens} = 4\,kg/cm^2$	Theory of elasticity for a fixed-ended beam with lateral load q.
8	12	2.5	l	Moment from plastic discs and no load in vertical direction	Equation (H.5) plus effect from lock discs.
9	129	26	l, h	Friction and moment from plastic discs plus $\sigma_{tens} = 0$ in vertical direction	Equation (H.5) plus effect from lock discs. Figure H.13.
10	211	43	l, h	$\sigma_{tens} = 4\,kg/cm^2$	Theory of elasticity for slabs.
11	867	177	l, h	All edges fixed	Yield line theory with moments according to equation (H.23).

H.10 □ Concluding remarks

The several different calculation methods for bending or combined bending and thrust on masonry walls that have been described in this chapter give good estimates of the load-carrying capacity of a masonry wall only when the assumed boundary and loading conditions are the same as in the practical case. The choice of method could be difficult in some instances; furthermore, some of the given numerical values of strength properties are scarcely verified by tests. As in any design situation, a complicated static system (as it appears in reality) has to be simplified to reflect only the fundamental behavior of the structure, and the less important secondary effects have to be abandoned.

To guide in the choice, the results from examples calculated by Sahlin and Hellers [H.9] are summarized in Table H.2. The ultimate lateral load a wall of a certain size can carry varies with a factor of 10, depending upon the boundary conditions. For example, the conditions shown in Fig. H.13 with the concentrated load replaced by a statically equivalent evenly distributed load result in a very low load compared to the load a wall supported along all sides can carry, according to the yield line theory. This of course also reflects the actual behavior of the structures since it is clear that a simple supported (at top and bottom) wall without tensile strength and with a low axial stress (1.6 kg/cm², 22 psi) must carry much less load than a wall with considerable tensile strength, supported along all four sides and actually functioning according to the yield line theory.

Reference for Chapter H

H.1 □ Hedstrom, R. O.: "Load Tests of Patterned Concrete Masonry Walls." ACI Journal, Proceedings, V. 57, p. 1265; PCA Development Department Bulletin D41, April, 1961.

H.2 □ Fishburn, C.: "Effect of Mortar Properties on Strength of Masonry." National Bureau of Standards, Monograph 36, Department of Commerce, Washington, D.C., November 20, 1961,

H.3 □ SCPRF. "Compressive, Transverse and Racking Strength Tests of Four-Inch Brick Walls." Research Report No. 9 of the Structural Clay Products Research Foundation, Geneva, Ill., 1965.

H.4 □ Benjamin and Williams: "The Behavior of One-Story Brick Shear Walls." Journal of the Structural Division, Proceedings of the American Society of Civil Engineers, Paper 1723, ST4, July, 1958.

H.5 □ Plummer, H., and Reardon, L.: "Principles of Brick Engineering, Handbook of Design." Structural Clay Products Institute, Washington, D.C., 1939.

H.6 □ Hallquist, Å.: "Vindtrykk på Skallmurer." (Wind Loads on Cavity Walls.) Tegl, Nos. 2 and 4. Hellström og Nordahl A/S, Oslo. (Journal from Teglverkenes Forskningsinstitutt, Forskningsveien 3b, Blindern, Norway), 1966.

H.7 □ Cox, F. W., and Ennenga, J. L.: "Transverse Strength of Concrete Block Walls." ACI Journal, Proceedings, Vol. 54, p. 951, May, 1958.

H.8 □ Timoshenko, S.: *Strength of Materials*, Part II. McGraw-Hill Book Co., New York, N.Y., 1941.

H.9 □ Sahlin, S., and Hellers, B. G.: "Transversalbelastning på Mellan Bjälklag Inspända Väggar utan Draghållfasthet." (Transversally Loaded Walls without Tensile Strength Restrained by Floor Slabs), Report 19, 1968. National Swedish Institute for Building Research, Stockholm, 1968.

H.10 □ Sahlin, S.: "Transversely Loaded Compression Members Made of Material Having No Tensile Strength." *IABSE*, Vol. 21, p. 243, Zürich, 1961.

H.11 □ Sahlin, S.: "Horisontella Punktlaster på Väggpelare av Låsfogad Ytongstav." (Horizontal Concentrated Loads on Wall Columns Made of Brick-Locked Ytong Blocks.) No. 48, Bulletins of the Division of Building Statics and Structural Engineering at the Royal Institute of Technology, Stockholm, 1963.

H.12 □ Grenley, D. G., Cattaneo, L. E., and Pfrang, E. O.: "The Effect of Edge Load on the Flexural Strength of Clay Masonry Systems Utilizing High-Strength Mortars," in *Designing, Engineering, and Constructing with Masonry Products*, edited by Dr. Franklin Johnson, Copyright © 1969 by Gulf Publishing Company, Houston, Texas. Used by permission.

H.13 □ Hellers, Bo-Göran: "Eccentrically Compressed Columns without Tensile Strength Subjected to Uniformly Distributed Lateral Loads." Report 35/67 from the National Swedish Institute for Building Research, Stockholm, 1967.

H.14 ☐ Timoshenko, S., and Young, D.: *Theory of Structures,* 2nd ed. McGraw-Hill, New York, 1965.

H.15 ☐ Young, D.: "Stresses in Eccentrically Loaded Steel Columns." Publication of the International Association of Bridge and Structural Engineering, Vol. 1, 1932, 507 pp.

H.16 ☐ American Institute of Steel Construction: "Manual of Steel Construction." Sixth Ed., AISC, New York, N. Y., 1964.

H.17 ☐ Broms and Viest: "Ultimate Strength Analysis of Long Restrained Reinforced Concrete Columns." Proceedings of the American Society of Civil Engineers, May, 1958.

H.18 ☐ Sahlin, S.: "Elementväggar Inbyggda Mellan Bjälklag —Spänningar, Deformationer, Kraftexcentriciteter och Bärförmåga. Del 1." (Wall Panels Built in Between Floors—Stresses, Deformations, Eccentricity of Forces, and Load-Bearing Capacity, Part 1.) Byggforskningen, Rapport 107, Stockholm, 1964.

H.19 ☐ Sahlin., S., and Jansson, S.: "Elementväggar Inbyggda Mellan Bjälklag—Spänningar, Deformationer, Kraftexcentriciteter och Bärförmåga, Del 2." (Wall Panels Built in Between Floors—Stresses, Deformations, Eccentricity of Forces, and Load-Bearing Capacity, Part 2.) Byggforskningen, Rapport 107, Stockholm, 1964.

H.20 ☐ McDowell, McKee, and Levin: "Arching Action Theory of Masonry Walls." Journal, Structural Division, Proceedings of the American Society of Civil Engineers, Paper 915, ST2, March, 1956.

H.21 ☐ Cohen and Laing: "Discussion of [H.20]" Journal, Structural Division, Proceedings of the American Society of Civil Engineers, Paper 1067, pp. 23–40, ST5, September, 1956.

H.22 ☐ Pettersson, O.: "Vindlastinverkan på Utfackningsväggar av Tegel." (Brick Masonry Curtain Walls Submitted to Wind Loading.) Division of Building Construction, Royal Institute of Technology, Stockholm, Sweden, Bulletin 4, Stockholm, 1959.

H.23 ☐ Nilsson, S.: "Vindbelastningar på Tegelväggar," Tegel No. 2, Stockholm, 1963.

H.24 ☐ Timoshenko, S., and Woinowsky-Krieger, S.: *Theory of Plates and Shells.* McGraw-Hill Book Co., New York, N.Y., 1959.

H.25 □ Bradshaw, R. E., and Entwisle, F. D.: "Wind Forces on Non-Loadbearing Brickwork Panels." Clay Products Technical Bureau, Technical Note B.1, No. 6, 1965.

H.26 □ Isaacs, D. V.: "A Second Interim Analysis of the Strength of Masonry Walls with Respect to Lateral Loading." Special Report No. 1, Commonwealth Experimental Building Station, Sydney, Australia, June, 1948.

H.27 □ Structural Clay Products Research Foundation: "Compressive and Transverse Strength Tests of Eight-Inch Brick Walls." Research Report No. 10, Geneva, Ill., 1966.

H.28 □ Structural Clay Products Research Foundation: "Compressive and Transverse Tests of Five-Inch Brick Walls." Research Report No. 8, Geneva, Ill., 1965.

H.29 □ Royen, N.: "Dimensionering av Murar med hänsyn tagen till Friktion och Vidhäftning." (Design of Masonry Walls, Accounting for Friction and Bond.) Häfte 9, Byggmästaren, Stockholm, 1936.

I ☐ Cavity walls

I.1 ☐ Introductory remarks

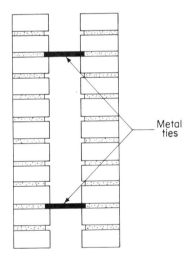

Figure I.1 ☐ Section of a cavity wall

A cavity wall (barrier wall, dual masonry wall) consists of two separate wythes tied together with metal ties (Fig. I.1). The two wythes can also be held together with bricks or blocks laid cross-wise (headers) between the two wythes at certain points. If the two wythes are close together and bonded with bricks closely spaced, the statical behavior of the wall approaches that of a solid wall since considerable shear forces can be transmitted between the wythes. Only the principal behavior of a typical cavity wall, with weak ties as shown in Fig. I.1 is treated here. The advantages of the arrangement are several. The continuous air barrier prevents moisture and heat (to some extent) to penetrate the wall. The void can be filled with insulating material such as glass wool to give better insulating performance. A cavity wall can be laid with both sides of facing bricks and thus requires no plastering. The load-carrying capacity, however, is mostly lower for a cavity wall than for a (well-bonded) two-wythe wall without cavity. The structural behavior of a cavity wall will be discussed in what follows.

I.2 ☐ One wythe loaded axially

A single-wythe wall is considered first (Fig. I.2). The moment at mid height due to the eccentric load P is

$$M = \frac{Pe}{\cos k\frac{h}{2}} \tag{I.1}$$

where $k = \sqrt{P/EI}$ and h is the wall height.

183

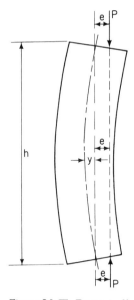

Figure I.2 □ Eccentrically loaded single-wythe masonry wall. Deflection = y.

For a wall with rectangular cross section, the tensile stress which governs cracking is

$$\sigma_t = \frac{-P}{A} + \frac{M}{W} \tag{I.2}$$

where A is the area of the cross section and W is the section modulus $(bd^2/6)$. Equations (I.1) and (I.2) then give

$$\sigma_t = \frac{-P}{bd} + \frac{6 \cdot Pe}{bd^2 \cos k\frac{h}{2}} \tag{I.3}$$

or

$$\sigma_t bd = P_t = -P + P\frac{6\frac{e}{d}}{\cos \frac{h}{2}\sqrt{\frac{P}{EI}}} = -P + P\frac{6\frac{e}{d}}{\cos \frac{\pi}{2}\sqrt{\frac{P}{P_{cr}}}} \tag{I.4}$$

if $P_{cr} = \pi^2 EI/h^2$.

The cracking load P_1 can be calculated from Eq. (I.4) and gives the result in implicit form:

$$P_t = P_1\left(\frac{6\frac{e}{d}}{\cos \frac{\pi}{2}\sqrt{\frac{P_1}{P_{cr}}}} - 1\right) \tag{I.5}$$

where $P_t = \sigma_t \cdot bd$; or with the notations $P_t/P_{cr} = p_t$, $P_1/P_{cr} = p_1$, and $e/d = e_r$,

$$p_t = p_1\left(\frac{6e_r}{\cos \frac{\pi}{2}\sqrt{p_1}} - 1\right) \tag{I.6}$$

Notations in Chapter I

A = area of cross section of wall	M_2 = moment on one wythe in a cavity (two-wythe) wall
b = length of wall	P = axial load on wall
d = thickness of wall	$P_{cr} = \pi^2 EI/h^2$
e = eccentricity of force	$P_t = bd\sigma_t$
e_r = relative eccentricity e/d	P_1 = cracking load of single-wythe wall
$E_k I_k$ = stiffness of a tie	P_2 = cracking load of a cavity (two-wythe) wall
EI = stiffness of wall	σ_t = bending tensile strengh (modulus of rupture) of wall
$k = \sqrt{P/EI}$	h = height of wall
M = bending moment acting on wall	$p_c = f'_m bd/P_{cr}$
	f'_m = compressive strength of masonry

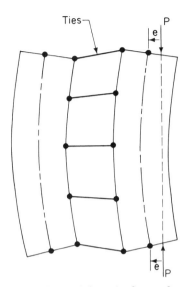

Figure I.3 □ Schematic figure of a cavity wall eccentrically loaded on one wythe. (The thickness to length ratio is exaggerated.) The ties are assumed to be linked to the wythes.

If a second identical wall is tied to the first one (Fig. I.1) in such a manner that both walls have the same curvature (which is an approximation; in reality, the deflections are equal at certain points), but the second wall does not carry any axial load, the buckling load will be doubled due to the doubled stiffness. The moment in the uncracked stage is then

$$M = \frac{Pe}{\cos \frac{\pi}{2} \sqrt{\frac{P}{2P_{cr}}}} \tag{I.7}$$

with the same notations as before (Fig. I.3).

The moment according to Eq. (I.7) is now shared equally by the two walls since they have the same curvature. The wythe without axial load is then loaded with the moment

$$M_2 = \frac{Pe}{2 \cos \frac{\pi}{2} \sqrt{\frac{P}{2P_{cr}}}} \tag{I.8}$$

only.

The maximum tensile stress at mid height is therefore

$$\sigma = \frac{M_2}{W} = \frac{6Pe}{bd^2 2 \cos \frac{\pi}{2} \sqrt{\frac{P}{2P_{cr}}}} \tag{I.9}$$

and with the same notations as before and the cracking load P_2, the following equation is derived:

$$P_t = \frac{3P_2 \frac{e}{d}}{\cos \frac{\pi}{2} \sqrt{\frac{P_2}{2P_{cr}}}} \tag{I.10}$$

or with $P_2/P_{cr} = p_2$

$$p_t = \frac{3p_2 e_r}{\cos \frac{\pi}{2} \sqrt{\frac{p_2}{2}}} \tag{I.11}$$

from which p_2 can be calculated.

From Eqs. (I.6) and (I.11), the values of p_1 and p_2 have been calculated for $p_t = 0.01$ and plotted as functions of the relative eccentricity e_r in Figs. I.4 and I.5. The ratio p_1/p_2 is over 10 when e/d is about 0.1, which means that for an eccentricity of 0.1 of the wall thickness the single-wythe wall has a cracking load which is more than 10 times the cracking load of a cavity wall, if the tensile strength of the wall is 1% of the buckling load. For very small eccentricities (less than 1.5% of the thick-

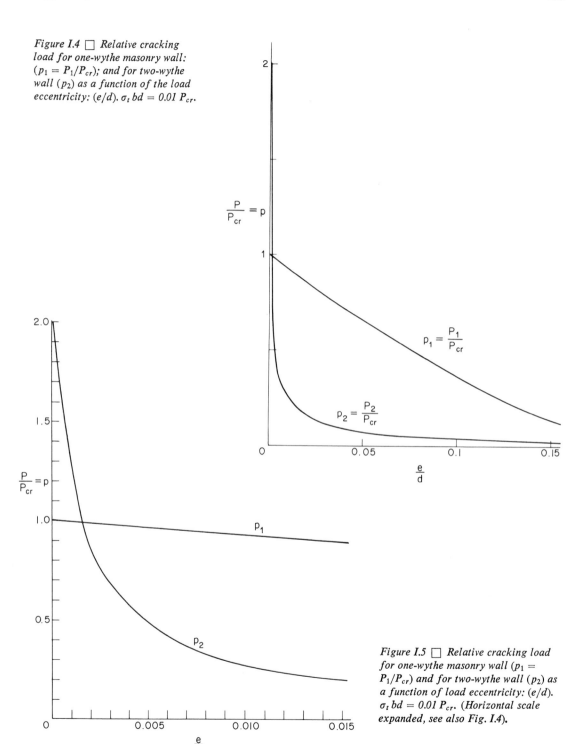

Figure I.4 ☐ Relative cracking load for one-wythe masonry wall: ($p_1 = P_1/P_{cr}$); and for two-wythe wall (p_2) as a function of the load eccentricity: (e/d). $\sigma_t\,bd = 0.01\,P_{cr}$.

Figure I.5 ☐ Relative cracking load for one-wythe masonry wall ($p_1 = P_1/P_{cr}$) and for two-wythe wall (p_2) as a function of load eccentricity: (e/d). $\sigma_t\,bd = 0.01\,P_{cr}$. (Horizontal scale expanded, see also Fig. I.4).

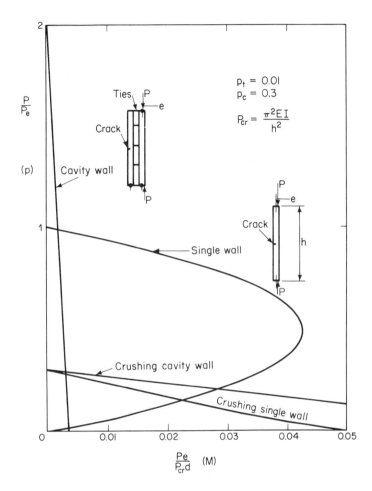

$p_t = 0.01$
$p_c = 0.3$

$P_{cr} = \dfrac{\pi^2 E I}{h^2}$

Figure I.6 ☐ Interaction diagram for the relative cracking load for single-wythe wall and cavity wall: $p_t = 0.01$. The relative crushing load has also been plotted for a crushing strength $f'_m \cdot bd = 0.3 p_{cr}$.

ness of one wythe), the cracking load for the cavity wall is higher than for the single-wythe wall. For concentric loading (which is almost impossible to accomplish in tests), the cavity wall has twice the strength of a single-wythe wall. For very large eccentricities (almost pure bending), the cracking load for the cavity wall again approaches a value which is twice as high as that for the single-wythe wall (not shown in Figs. I.4 and I.5).

For large eccentricities the phenomenon is better displayed in an interaction diagram (Fig. I.6). In Fig. I.6 the compressive strength $f'_m bd = p_c \cdot P_{cr}$ has also been considered. The relative value p_c has been assumed to be 0.3. For only two small areas in Fig. I.6 the cavity wall has higher cracking or crushing load

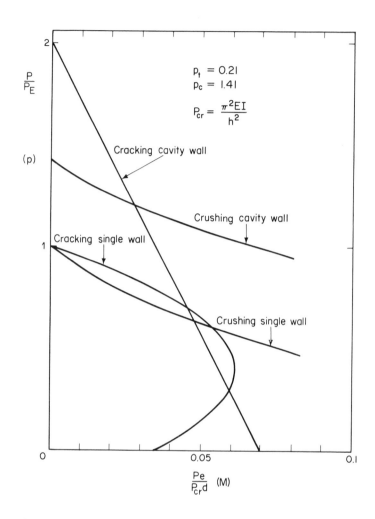

Figure I.7 □ Interaction diagram for the relative cracking load and for the relative crushing load for single-wythe wall and cavity wall: $p_t = 0.21$; $p_c = 1.41$.

than the single-wythe wall. If p_c is assumed to be more than 2, the cracking loads are lower than the crushing loads for both types of walls for most loading combinations. For an intermediate level ($p_t = 0.21$; $p_c = 1.41$), an interaction diagram is shown in Fig. I.7.

It can be seen from the figures that the cavity wall has a very low cracking moment (for low tensile strengths) compared with the single-wythe wall. This is especially true when the axial load is not too low. The reason is that the unloaded wythe of the cavity wall does not benefit from the axial compression that counteracts the bending tensile stresses, in the case of the single wall, or in the loaded wythe in the cavity wall.

It was assumed above that both wall wythes had the same thickness and the same curvature. Furthermore, an axial load was acting parallel to the wall axes and no lateral loads were present. When one or more of these assumptions are invalid, the given equations must be properly adjusted or even replaced.

I.3 ☐ Effect of ties

As shown in Fig. I.3, the ties were assumed to be pin-ended; i.e., no shear or moment is transmitted from one wythe to the other by the ties. Normally the stiffness of the ties is very small compared with the stiffness of the masonry wythes. For stiff ties closely spaced, a theory derived for sandwich panels could be applied. This theory has been presented by Holmberg and Plem [I.1]. For weak ties the problem has been studied for a basic case. (See Fig. I.8.) The tie transmits the lateral force H (the axial force in the tie), the axial force V (the shear force in the tie), and the moment M_0 to the wall end (Fig. I.9). After some lengthy derivations [I.2], formulas for stresses at different points (E, C, D, and F in Fig. I.8) were obtained and evaluated on a CDC 3200 computer to give cracking loads for the original wall system which normally cracks first at point E (Fig. I.8). The cracking load was computed for a practical value of the tie's stiffness

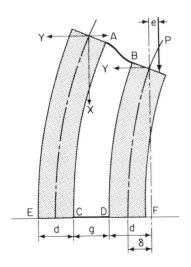

Figure I.8 ☐ Loading case for a study of the effect of ties on the cracking load of a cavity wall [I.2].

$$\mu = \frac{6E_k I_k}{EAg^2} \cdot \frac{d+g}{g} = 3.85 \cdot 10^{-5}$$

$$\gamma = \frac{6E_k I_k}{EI} \cdot \left(\frac{d+g}{g}\right)^2 = 1.083 \cdot 10^{-3}$$

as well as for the stiffness μ and $\gamma = 0$. $E_k I_k$ refers to the ties and EI to one single wythe, d is the wall thickness and g is the cavity thickness. The influence of the tie stiffness was hardly noticeable.

The cracking load was also computed for a second basic case in which it was assumed that the axially unloaded wythe had already cracked (Fig. I.10). The tensile crack would, in the latter case, appear at point D. Even for the situation shown in Fig. I.10, the cracking load was calculated for ties with a practical stiffness as well as for laterally completely flexible ties.

For the loading case shown in Fig. I.10, the cracking load of the axially loaded wythe (after that the unloaded wythe was cracked) was influenced very little, less than 1%, by the increase

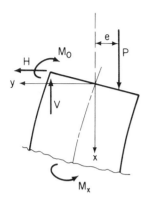

Figure I.9 □ Loading conditions at the end of the axially loaded wythe of a cavity wall. See Fig. I.8 [I.2].

in tie stiffness. The data were for the studied cases: masonry tensile strength 10 kg/cm² (142 psi)

$$\mu = 0 \quad \text{or} \quad 0.000385$$

$$v = 0 \quad \text{or} \quad 0.001083$$

$$g = 10 \text{ cm} = d$$

$$L = 125 \text{ cm}$$

$$0 = \; \leq e \leq 4.5d$$

A calculation of the crushing load was also made for the same case, and again the differences were small (less than 1%) when the axial load acted inside the section and somewhat higher for higher eccentricities, but the approximations in the theoretical model probably also influenced the results. The crushing load was calculated for a compressive strength of 100kg/cm²(1422 psi).

Although the stiffness of the ties was beneficial to the load-carrying capacity of the walls in the cases studied, the influence was mostly so small that the effect can be neglected for practical purposes. The effect from the ties will probably depend strongly on the support conditions for the unloaded wythe, and it can easily be conceived that the cracked unloaded wythe acts as a load attached by the ties to the loaded wythe.

Figure I.10 □ Loading case for a study of the effect of ties on the cracking load of a cavity wall. The unloaded wythe is cracked [I.2].

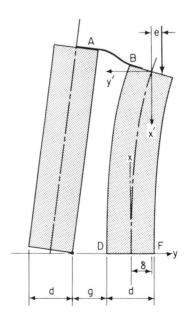

I.4 □ Both wythes loaded

If both wythes are loaded simultaneously with equal loads and equal eccentricities, the total axial cracking load of the assembly would be two times the axial cracking load for a single-wythe wall because then both wythes would be identical in all respects, even without ties. For cases where one wall is lightly loaded and the other heavily loaded, the results will depend upon the directions and magnitudes of the eccentricities, but it is probable that the resulting cracking load is somewhere between the load calculated according to Section I.2 and the cracking load obtained for a cavity wall with identical loading on both wythes.

References for Chapter I

I.1 □ Holmberg, A., and Plem, E.: "Behavior of Load-Bearing Sandwich-Type Structures." National Swedish Institute for Building Research, Handlingar, No. 49, 1965.

I.2 □ Sahlin, S., and Sundquist, H.: "Principstudium av kramlors effect vid axial-belastning av dubbelvägg av tegel" (Unpublished manuscript, 1967).

J □ Cracking

J.1 □ Introductory remarks

Cracking can be an annoying phenomenon in masonry structures. The main cause is the differential movements of different building parts (or materials). A difference in stress of, say, 100 psi in two adjacent masonry walls would, at the top of a 10-story building, add up to about $\frac{1}{2}$ in. differential vertical movements. A 100-ft long masonry wall of severely shrinking materials would shorten about $\frac{1}{2}$ in. if unrestrained. If the walls mentioned above were restrained by other building elements, cracking would probably occur. Deflection and shrinkage of concrete slabs resting on the walls can also cause cracking. Problems of this kind are dealt with in what follows; we begin with the effect of slab deflections and then proceed to differential strains from loads and temperature differences. Tests on restrained shrinking walls are then described, and the chapter is concluded with a section on experiences from practice.

J.2 □ Effects of deflections of slabs on the cracking of external walls

Figure J.1 from [J.1] shows a mechanism of crack development which may take place when a floor slab resting on a wall is loaded. The load q is assumed to be evenly distributed along the span L, and no other loads are applied. This means that the situation described in Fig. J.1 usually will appear only at the top floor slab in a building where no loads from walls on the top of the slab edge exist. Due to the rotation of the end of the slab, a crack will develop between the slab end and the wall. The situation is somewhat similar to that at the support of an intermediate slab (Fig. J.2), but the cracking risk decreases as the force in the wall (and the restraining moment on the slab end) increases. Therefore, only the situation shown in Fig. J.1 will be dealt with

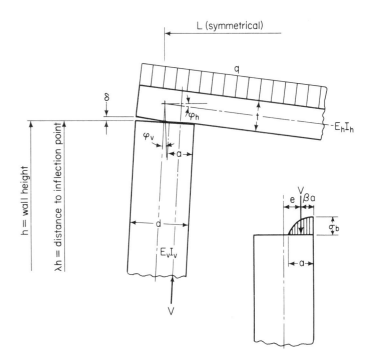

Figure J.1 ☐ *Detail of connection between floor slab and wall after loading. Top story crack width shown as δ [J.1].*

Figure J.2 ☐ *Detail of connection between floor slab and wall after loading, intermediate story. Mechanism of cracking is shown in principle.*

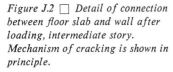

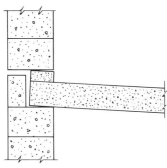

in detail. An estimate of the crack width δ (Fig. J.1) can be obtained from Eq. (J.9) derived below.

The floor slab (Fig. J.1) with the flexural rigidity $E_h I_h$ is supported by the wall which has the rigidity $E_v I_v$, the thickness d, and the height h. The distance from the slab to the nearest inflection point (zero moment point) in the wall is λh. Normally λ has a value between 0.5 and 0.7. The angle of rotation of the end of the slab is denoted by φ_h. Finally, the wall and the slab are in contact along a distance a. With the assumptions mentioned, the crack width δ is obtained from

$$\delta = (d - a)(\varphi_h - \varphi_v) \tag{J.1}$$

or

$$\delta = (d - a)\theta$$

(See Chapter F for a more detailed discussion of θ.)

The stress distribution over the contact distance a is unknown but is assumed to have the approximate shape shown in Fig. J.1. By introduction of a shape factor α and an ultimate stress σ_b, the distance a is calculated to be

$$a = \frac{V}{\alpha \sigma_b b} \tag{J.2}$$

Notations in Chapter J

a = distance of contact between wall and slab	α = shape factor
b = breadth (length) of wall	βa = the distance from the wall edge to the re-
$C = 2E_h I_h \lambda h d / E_v I_v L^2$	sulting force over the area of contact be-
d = thickness of wall	tween slab and wall
e = eccentricity of axial force in the wall	δ = crack width
$E_h I_h$ = rigidity of horizontal member	ε = strain
$E_v I_v$ = rigidity of vertical member	θ = angular rotation of joint
f' = strength	λh = distance from slab to inflexion point below
$f(V, e)$ = a function of V and e	the slab
h = height of wall	σ = stress
$K = L^2 d / 12 E_h I_h$	σ_b = ultimate stress of wall material
L = span of slab	σ_{perm} = permissible stress
q = unit distributed load on slab	φ_h = angular end rotation of horizontal member
V = axial vertical force in wall	φ_v = angular end rotation of vertical member
$V_E = \pi^2 E_v I_v / (\lambda h)^2$	

where b is the length of the wall and V the vertical load in the wall. Furthermore, for a symmetrical loading case,

$$\varphi_h = \frac{qbL^3}{24E_h I_h} \tag{J.3}$$

if $e \ll L$ and $d \ll L$.

The end rotation of the wall can be calculated according to Section E.2e or F.8.

Generally,

$$\varphi_v = \frac{Ve\lambda h}{3E_v I_v} f(V, e) \tag{J.4}$$

From Fig. J.1,

$$e = \frac{d}{2} - \beta a \tag{J.5}$$

if βa expresses the location of the resulting force measured from the edge of the wall. The stress distribution over the contact area between the slab and the wall is assumed to be somewhere between triangular and rectangular. Hence

$$\left. \begin{array}{c} \dfrac{1}{3} \leq \beta \leq \dfrac{1}{2} \\[2mm] 1 \leq \dfrac{2\beta}{\alpha} \leq \dfrac{4}{3} \\[2mm] \dfrac{1}{2} \leq \alpha \leq 1 \end{array} \right\} \tag{J.6}$$

Furthermore, it is assumed that

$$0 \le V \le \frac{1}{3} V_E = \frac{\pi^2 E_v I_v}{3(\lambda h)^2} \tag{J.7}$$

and that

$$0 \le V \le 0.1\sigma_b bd \tag{J.8}$$

Finally, an upper and lower boundary for δ is calculated and the approximations of α, β, and $f(V, e)$ are accepted if $\delta_{\min}/\delta_{\max} \ge 0.5$. These approximations lead to the following final expression:

$$\delta = V \frac{L^2 d}{12 E_h I_h} \left(1 - \frac{2 E_h I_h \lambda h d}{E_v I_v L^2}\right) = V \cdot K(1 - C) \tag{J.9}$$

which is valid for $C < 0.7$.

To sum up, the calculation procedure is as follows: Check Eqs. (J.7) and (J.8). Calculate K and C. Then δ can be calculated from Eq. (J.9). [Eq. (J.9) is valid only if $C < 0.7$.] If the crack width is unacceptable (that is, if it easily can be noticed, for example) a change in slab stiffness or an external cover of the crack can be necessary.

If the boundary conditions of the slab are different from those assumed above, a fictitious value for the span length L_{id} can be inserted in Eq. (J.9) so that Eq. (J.3) is satisfied; i.e., choose the fictitious length so that the fictitious simply supported beam, Eq. (J.3), has the same end rotation as the actual slab.

J.3 □ Effect of differential strain from stress and temperature changes in walls and columns

Consider a building with a cross section as shown in Fig. J.3. Assume that the central wall is loaded to a stress of 200 psi, while the external walls are loaded to 100 psi only. Suppose that the (load-bearing) walls are built of different materials so that the external walls have a modulus of elasticity of $3 \cdot 10^6$ psi and the central wall, $0.5 \cdot 10^6$ psi. The strain due to load in the central wall is therefore

$$\varepsilon = \frac{200}{0.5 \cdot 10^6} = 0.4 \cdot 10^{-3}$$

and in the external walls

$$\varepsilon = \frac{100}{3 \cdot 10^6} = 0.33 \cdot 10^{-4}$$

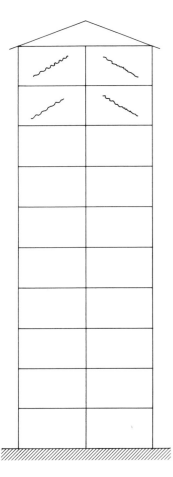

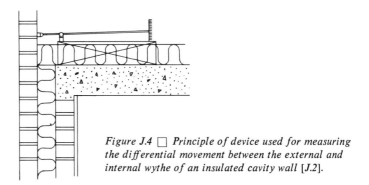

Figure J.4 ☐ *Principle of device used for measuring the differential movement between the external and internal wythe of an insulated cavity wall [J.2].*

At the top of a 10-story building of 100-ft height, the differential strain would add up to a differential deformation of $(0.4 \cdot 10^{-3} - 0.33 \cdot 10^{-4})100 \cdot 12 \approx 0.44$ in. $= 11$ mm if no restraint is present. Normally the differential deformation is at least to some extent prevented by the partitioning walls. If these are stiff enough, they will crack as indicated in Fig. J.3. In reality the external walls are also subjected to fluctuating temperatures and, therefore, thermal deformations have to be added to the stress dependent deformations.

Measurements by Nevander [J.2] show about 0.4 mm (0.16 in.) daily movements due to temperature changes in the external wythe of a cavity wall on a three-story building. The wall was 9 m (30 ft) high and the readings were taken at the attic as the difference in movement between the outer and inner wythe of

Figure J.3 ☐ *(Above) Cross section of building. The central wall is assumed to be compressed more than the external walls (See discussion in text). Cracking in cross walls occurs due to differential movements.*

Figure J.5 ☐ *(Right) Readings of movements and temperatures from the measurements described in Fig. J.4 [J.2].*

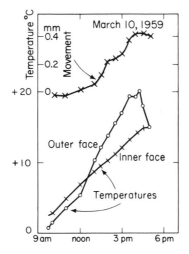

the cavity wall (Fig. J.4). The wall temperatures and the movements during a sunny early spring day are shown in Fig. J.5. The sun reached the wall at about noontime.

These examples of calculations and measurements show that, to avoid cracking, caution should be exercised when choosing material and stress levels for walls in higher buildings at least.

J.4 □ Cracking due to shrinkage and creep

J.4a: Introductory remarks

A masonry wall is normally restrained to some extent along its edges or at certain points. Differential shrinkage between the masonry and the restraining media will therefore build up stresses in the masonry. Due to creep, such stresses are released to some extent in certain cases. In other cases the creep can be a source of differential movements and accompanying stresses. The basic phenomena of shrinkage and creep are discussed first in this section; then the deformations of masonry under long-term loading are treated. Finally, the stresses, strains, and crack formations in a restrained shrinking wall are discussed.

J.4b: Shrinkage

Most masonry types shrink with time. The minimum shrinkage that has to be expected is the part contributed by the shrinkage of the mortar joints. The joints do not (or seldom) exceed 20% of the wall height and about 10% of the length. (A typical value is 15% for clay brick masonry and 5% for concrete block unit masonry in the vertical direction.) The major part of any considerable masonry shrinkage is therefore usually due to the shrinkage of the masonry units. The potential shrinkage of the units is strongly dependent on the manufacturing process and the pretreatment of the units (steam curing, autoclaving, burning, storing for a time, drying, etc.).

Burnt clay bricks have practically no shrinkage (except for reversible expansions or contractions due to temperature and moisture content changes).

Light weight cellular concrete (of Siporex or Ytong type) is usually delivered from the factory with a moisture content of 30 to 40% (by weight). Since the moisture content stabilizes around 5% after long time exposure in normal climatic conditions (-20 to $+20°C$ and 30 to 70% RH), drying shrinkage

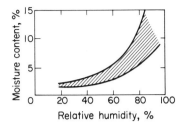

Figure J.6 □ Moisture content of light weight cellular concrete as a function of relative humidity at normal air temperature [J.3].

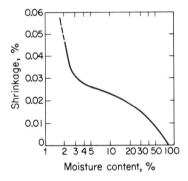

Figure J.7 □ Shrinkage of light weight cellular concrete (after 3 days storage in water) to equilibrium in air of room temperature and 43% RH [J.3].

takes place. The potential drying shrinkage can be estimated from Figs. J.6 and J.7, taken from [J.3]. Figure J.6 gives the moisture content related to the relative humidity, and Fig. J.7 gives the shrinkage for a change in moisture content. In a building with a mixture of thin and thick walls, the difference in drying time can cause considerable differential shrinkage.

The shrinkage of *concrete masonry blocks* varies within rather wide limits. It has been shown by standard tests on laboratory specimens made by ACI Committee 716 [J.4] that the type of aggregate used in the concrete masonry blocks affects the magnitude of shrinkage (Fig. A.9). The same can also be seen from tests by Kuenning, William, and Carlson [J.5] (compiled in Fig. J.8). In this test series [J.5], it was shown that shrinkage of blocks cured at high temperatures (110°F, 15 days total curing) was about 50 to 70% of that measured on low temperature cured blocks, except for pumice plus cement III blocks, for which the percentage was 35.

The shrinkage of concrete blocks steam cured at atmospheric pressure is reduced by precarbonation, as shown by Shiedeler [J.6]. Precarbonation also reduces subsequent length changes due to wetting and drying. Carbonation of autoclaved blocks, however, gives an increased expansion during immersion for slag and shale blocks. A detailed discussion of the effect of changes in the

Figure J.8 □ Shrinkage of concrete masonry units made of different aggregates, moist cured blocks [J.5].

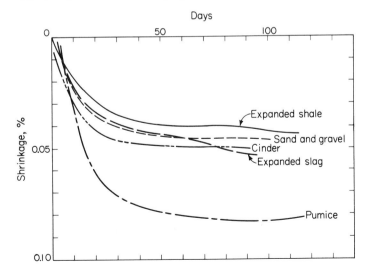

Figure J.9 ☐ *(Right) Potential shrinkage of block with moisture loss and carbonation during long time exposure to air at various relative humidities [J.8]. If it is assumed that moisture from the mortar increases the humidity of block initially at 70 to 85% (15% increase) and that of block initially at 50 to 70% (20% increase), the indicated moisture shrinkages A and B would occur.*

composition of aggregate and the treatment of concrete unit can be found in Shiedeler [J.6] and Verbeck [J.7]. Menzel [J.8] has compiled data from Shiedeler and Verbeck and plotted the potential shrinkage of concrete blocks with moisture loss and carbonation during lengthy exposure to air at various relative humidities (Fig. J.9). From other sources, Menzel concludes that, in general, blocks cured in an autoclave in steam at 120 psi gauge pressure for 8 hours exhibit about 50% of the shrinkage of blocks cured in atmospheric pressure. In Fig. J.9, diagrams adopted from Menzel [J.8] show the potential shrinkage of concrete blocks steam cured at atmospheric pressure, and estimated values for blocks steam cured at 120 psi. Somewhat less influence due to autoclaving was found by Hedstrom, Litvin, and Hanson [J.9], as seen in Table J.1. Effect of aggregate changes can also be seen in Table J.1.

The *mortar shrinkage* can be expected to be of the same magnitude as for concrete, i.e., 0.01 to 0.08%, the lowest figure for low flow–low water to cement ratio and the higher for high flow–high water to cement ratio concrete. For lime mortars the figures are considerably higher, especially at an early age. See Fig. J.10 from [J.10].

The shrinkage of masonry units is also often strongly related to the *relative humidity* (RH). (An increase in RH due to the yearly variations can thus also cause expansion over certain periods.) The effect is of the type shown in Figs. J.6 and J.7 for several materials.

Figure J.10 ☐ *Shrinkage of some mortar types up to 8 days of Age. (KC 35/65 is the weight ratio of lime to cement, etc.) [J.10].*

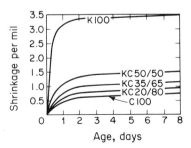

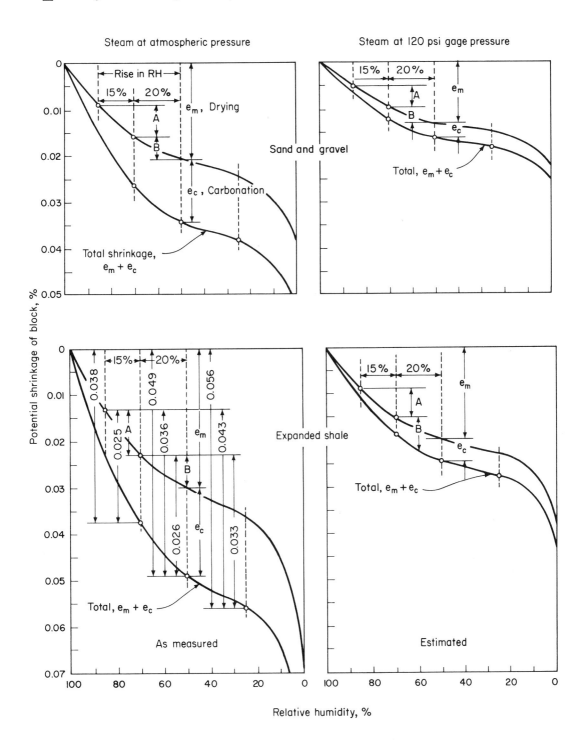

Steam at atmospheric pressure

Steam at 120 psi gage pressure

Sand and gravel

Expanded shale

As measured

Estimated

Potential shrinkage of block, %

Relative humidity, %

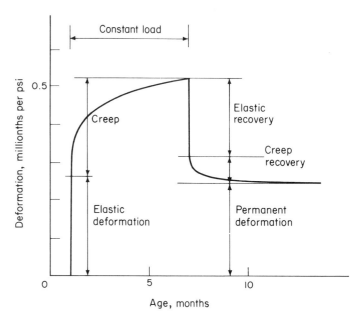

Figure J.11 □ *Creep and shrinkage of concrete under loading, storing, and unloading: principal behavior.*

J.4c: Creep

When a concrete masonry block is loaded, a nearly elastic strain is observed (for low stress levels). The strain, however, increases with time, even under constant stress. This time-dependent strain increase is called *creep* (or plastic flow), and is sometimes of the same magnitude as the elastic strain. Loading at an early age, drying, and high water to cement ratio usually increase the creep for masonry units of cementitious materials. Completely wet or dry concrete creeps less; in general, creep decreases with the age of the concrete.

Table J.1 □ *Shrinkage of hollow concrete block from saturation to 50% RH [J.9]. Some of the materials were selected to give high shrinkage because the main purpose of the investigation was to study shrinkage cracking.*

Type of block	Compressive strength (gross area)		Drying shrinkage saturation to 50% RH
	kg/cm²	psi	%
Expanded shale			
Atmospheric steam cured	128	1820	0.0392
Autoclaved	125	1780	0.0328
Sand and Gravel			
Atmospheric steam cured	169	2400	0.0308
Autoclaved	163	2320	0.0260
Expanded slag			0.0482
Granulated slag	103	1460	0.0670

Creep is approximately directly proportional to the unit stress for low stress levels, and a loading, storing, and unloading process could be represented in principle by Fig. J.11. The creep of mortar follows the general pattern outlined for concrete. The creep phenomena of light weight cellular concrete are also similar to those of normal weight concrete. The creep of burnt clay bricks is of no significance under normal conditions.

J.4d : Masonry under long-term loading

Nylander and Ericsson [J.11] tested clay brick masonry columns, light weight cellular concrete masonry columns, and solid concrete brick masonry columns for a period of 400 days, with load (stress level around the permissible stress) and without load, and reported shrinkage and creep for different types of masonry laid with three different types of masonry units. In addition, different types of mortar were also tested as shown in Table J.2. The bricks had a compressive strength of 98 kg/cm² (1380 psi). The compressive strength of the light weight cellular concrete blocks was 39 kg/cm² (555 psi) and of the concrete bricks, 159 kg/cm² (2260 psi). The different tests are listed in Table J.2. Nylander and Ericsson's test shows that the time-dependent deformations for clay brick masonry are small, that untreated concrete block masonry has considerable shrinkage, and that a light weight cellular concrete masonry wall has considerable response to changes in RH and moisture content. See Figs. J.12, J.13, and J.14 for details. The RH under the test period is shown in Fig. J.15.

	Mortar proportions by weight		
	Cement–lime–sand		Lime–sand
Masonry units	1 : 1 : 12 $f' = 19\,\mathrm{kg/cm^2}$ (270 psi)	2 : 1 : 16 $f' = 54\,\mathrm{kg/cm^2}$ (770 psi)	1 : 7 $f' = 6.4\,\mathrm{kg/cm^2}$ (91 psi)
Brick $\sigma_{\mathrm{perm}} = 8\,\mathrm{kg/cm^2}$ (114 psi)	2 + 2C + 1S	2 + 1S	2 + 2C + 1S
Light weight cellular concrete $\sigma_{\mathrm{perm}} = 3\,\mathrm{kg/cm^2}$ (43 psi)	2 + 3C + 1S	2 + 1S	2 + 2C + 1S
Concrete brick $\sigma_{\mathrm{perm}} = 6\,\mathrm{kg/cm^2}$ (86 psi)	1 + 2C + 1S	2 + 1S	2 + 2C + 1S

Table J.2 ☐ Tests on masonry columns by Nylander and Ericsson[J.11]. Notation: 2 + 2C + 1S = two short-term strength, two creep test, and one shrinkage test specimens.

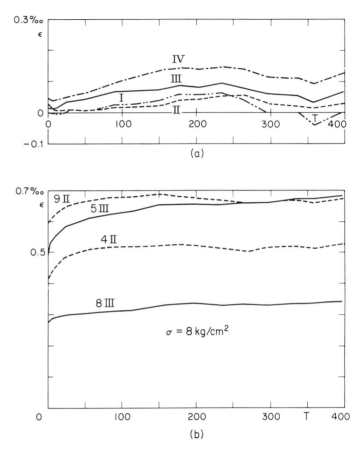

Figure J.12 ☐ Long-time deformations of brick piers. a. Shortening of unloaded piers. b. Difference in shortening between loaded and unloaded piers at σ = σ$_{perm}$ = 8 kg/cm²; T = time, days; ε = shortening, per mil (− = elongation); I: single Brick; II: brick pier, lime mortar; III: brick pier, with mortar of 2 parts lime to 1 part cement to 12 parts sand by volume; IV: brick pier, with mortar of 1 part lime to 1 part cement to 8 parts sand by volume. Prior to the time 0, the test specimens were stored for some 4 weeks in air at a relative humidity of about 70%. Arabic numerals give the designations of the loaded test specimens [J.11].

The creep for the different types of masonry is shown in Fig. J.16, but the number of tests does not allow a detailed separation of factors affecting the total creep. The results, however, give a basis for estimating the long-term deformations at permissible stress of masonry of the types tested.

The creep for early loading of brick masonry is demonstrated in Fig. J.17 from [J.12], which shows results from tests simulating the loading process for a wall in the lowermost part of a building which is successively loaded by the addition of stories higher up in the building. Two types of walls were tested, one wall was built with lime mortar and one with lime–cement mortar. The load was also removed after testing to show the recovery of the deformations.

Figure J.13 ☐ Long-time deformations of light weight concrete piers. a. Shortening of nonloaded piers. b. Difference in shortening between loaded and nonloaded piers at σ = σ_perm = 3 kg/cm². For notation and other particulars, see Fig. J.12 [J.11].

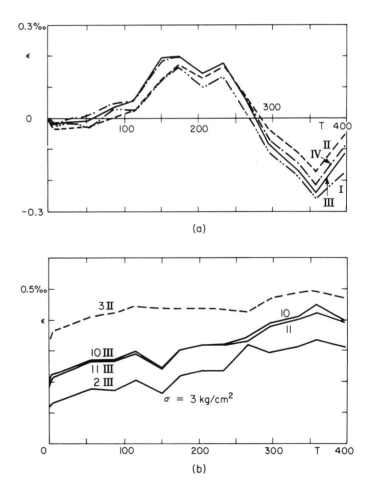

(a)

(b)

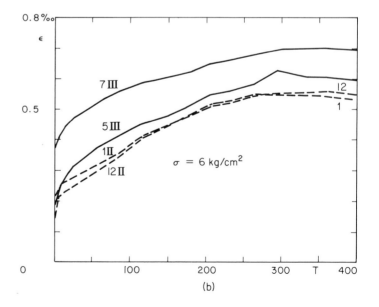

Figure J.14 ☐ Long-time deformations of concrete brick piers. a. Shortening of nonloaded piers. b. Difference in shortening between loaded and nonloaded piers at σ = σ_perm = 6 kg/cm². For notation and other particulars, see Fig. J.12 [J.11].

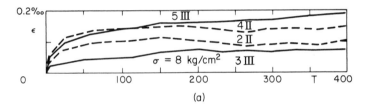

Figure J.15 □ Variation in the relative humidity of the air in the test room during the test period (Figs. J.12, J.13, J.14) [J.11].

Figure J.16 □ Creep of piers calculated in terms of the difference in longtime deformation between loaded (constant load) and nonloaded test specimens [J.11].
a. Brick piers
  $\sigma = \sigma_{perm} = 8\,kg/cm^2$
b. Light weight concrete piers
  $\sigma = \sigma_{perm} = 3\,kg/cm^2$
c. Concrete block piers
  $\sigma = \sigma_{perm} = 6\,kg/cm^2$

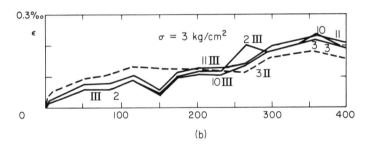

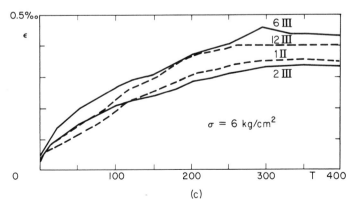

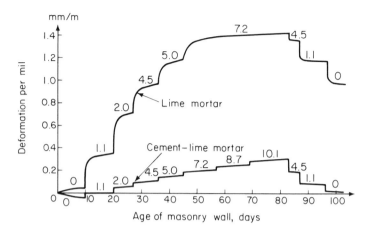

Figure J.17 □ *Compression of clay brick masonry walls subjected to stepwise increasing loads. The readings began 1 to 2 h after the walls were laid. Joint thickness, 12 mm ($\frac{1}{2}$ in.); modulus of elasticity of bricks 85,000 kg/cm² (1,210,000 psi); brick strength 190 kg/cm² (2700 psi). The figures on the curves indicate the compressive stress in kg/cm². The tests were run under laboratory conditions. Lime mortar was used for one wall and lime-cement mortar for the other [J.12].*

J.4e : Stresses and strains caused by shrinkage in restrained walls

Concrete block masonry walls of granulated slag, three-core 8 × 8 × 16 in. blocks and different types of mortars were tested in (1) tension, (2) complete restrain and drying in 50% RH for about 50 days plus subsequent tension until failure, and (3) complete restrain and drying from a moist condition until failure in 50% RH by Hedstrom, Litvin, and Hanson [J.9]. The walls were all loaded with a constant 20 psi stress perpendicular to the bed joints to simulate vertical load in the wall.

The stress–strain diagrams recorded for walls tensioned parallel with the bed joints are shown in Fig. J.18. The walls were laid in *N* mortar (of masonry cement) with 113% flow, 65 kg/cm² (925 psi) compressive and 12 kg/cm² (175 psi) tensile strength. The wall length is made up of about 2.5% mortar joints and 97.5% block lengths; but of the total elongation of the wall more than $\frac{2}{3}$ occurred in form of elongations or cracks in the vertical mortar joints (Fig. J.18).

Another series of tests was run on essentially the same type of walls as presented in Fig. J.18. In this case the walls were

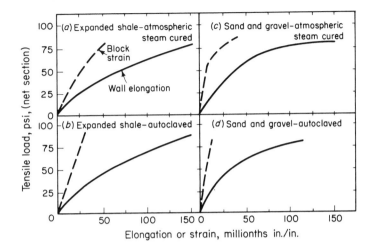

Figure J.18 □ Tensile
deformations of masonry walls
erected and cured at 50% RH,
series T [J.9].

erected and stored 28 days in 75% RH. Thereafter, the walls
were prevented from changing length and dried to equilibrium
in 50% RH. The resulting stresses were 45 to 65 psi, as shown
in Table J.3. The walls did not fail and were subsequently
slowly stressed to failure at 6.0 to 6.7 kg/cm² (85 to 95 psi)
which was slightly higher than for the directly loaded specimens
reported in Fig. J.18. The effect of the restraining on the
shrinkage of the blocks and the walls as a whole is shown in
Fig. J.19.

A third test series was laid of highly shrinking blocks and
stored in 50% RH, thereafter, stored for 3 weeks in a moist room,
restrained, and dried in 50% RH. Walls of eight different mor-

Figure J.19 □ Shrinkage
deformations of walls and block,
series II [J.9].

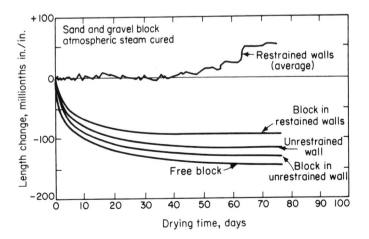

Table J.3 □ Loads on walls at moisture equilibrium [J. 9].

Type of unit	Tensile loads on walls at equilibrium, psi	
	Average	Range
Expanded shale—atmospheric steam cured	65	±5
Expanded shale—autoclaved	45	±5
Sand and gravel—atmospheric steam cured	65	0
Sand and gravel—autoclaved	55	±5

tars were tested. The restraining force increased as the shrinkage progressed up to the point when the tensile bond strength of the vertical joints was exceeded. When the vertical joints failed, the elongations across the joints increased rapidly and continued at a rate greater than if the joints had been closed. All walls in this test series cracked through a vertical joint and across the adjacent blocks, rather than along a zigzag line breaking the horizontal joints in shear. The shear strength of all mortars (*O, N, S, M*) was high in this test series, probably due to the beneficial effect on the hardening procedure that resulted from the moist storing of the walls.

The walls with the weakest mortars proved to be the strongest since the weakest mortars accommodated larger deformations before failure and tended to decrease stress concentrations. The walls with *O* mortars produced walls with a tensile strength of about 43% of the block tensile strength and the *M* mortars only 30%. Since the walls with *O* mortars had a lower modulus of extension (the joints deformed more) and were stronger, these walls accommodated about twice as much shrinkage as the walls with the *M* mortars.

The block shrinkages in unrestrained and restrained walls are shown in Fig. J.20 for four different types of mortars [J.9]. (The free block shrinkage from saturation to 50% RH was 0.00067 for these high shrinkage blocks. The wall shrinkage is also indicated in Fig. J.20. The differences in deformations of blocks and walls are accommodated by the joints or, in the restrained case, by stresses and strains in the blocks and the joints.

The joint openings for different equivalent block strinkages as well as failure elongations for different mortar types are summarized in Fig. J.21 [J.9]. With no restraints of the blocks from the mortar, the vertical joint openings would be the distance calculated from the shrinkage of the blocks. This gives the straight line in Fig. J.21. With mortars stronger than the blocks, no joint openings would occur and failure would occur when the extensibility of the blocks was reached (0.00010 to 0.00015).

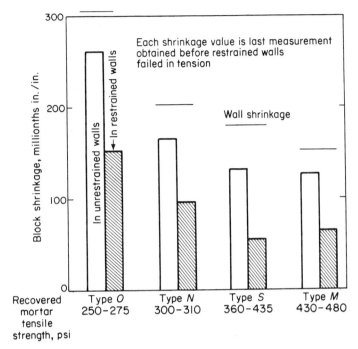

Figure J.20 ☐ (Right) Block shrinkage in unrestrained and restrained walls erected with different types of mortar, series III [J.9].

Each shrinkage value is last measurement obtained before restrained walls failed in tension

Wall shrinkage

In unrestrained walls
In restrained walls

Block shrinkage, millionths in./in.

| Recovered mortar tensile strength, psi | Type *O* 250–275 | Type *N* 300–310 | Type *S* 360–435 | Type *M* 430–480 |

Figure J.21 ☐ (Below) Average vertical joint opening in fully restrained walls, series III. Directions for use: The block shrinkage is entered on the horizontal axis and by using the curve in the diagram the vertical joint openings are read on the vertical axis. The letters on the curve represent cracking of a wall built of the indicated type of mortar, 100% restraint [J.9] (Based on test data from drying shrinkage from high moisture content to equilibrium at 50% RH).

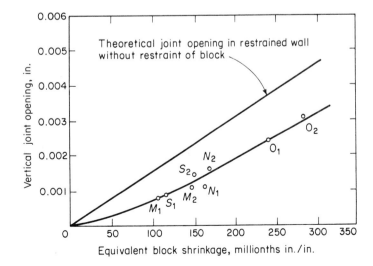

Theoretical joint opening in restrained wall without restraint of block

Vertical joint opening, in.

Equivalent block shrinkage, millionths in./in.

The curve of Fig. J.21 represents the potential shrinkage–joint-opening relation for walls with 100% restraint. The points indicate the failure conditions for each of the eight types of mortars used. Walls with less restraint would give curves below the one shown in Fig. J.21.

The related tests [J.9] show that shrinkage is accommodated to a greater extent by a wall with weak mortar than by a wall with stronger mortar. Whether failure occurs depends on the potential shrinkage of the blocks (Fig. J.21). Mortars containing masonry cement accommodated somewhat greater block shrinkage. A reduction in restraint resulted in greater accommodation of block shrinkage. Some potential block shrinkage (low shrinking type of blocks installed in dry condition) resulted also in greater accommodation of block shrinkage since a low final shrinkage also implies a low rate of shrinkage, allowing more time for shrinkage stresses to be relaxed by creep in mortar and concrete.

As soon as the bond strength between the vertical mortar joints and the blocks is exceeded, cracks will open in these joints. The joint openings are 0.001 to 0.003 in. wide prior to failure [the wall is divided in two or more parts by continuous crack(s)] of the wall. The restraint can be reduced by introduction of control joints. Cracking may develop in shear through the vertical and horizontal mortar joints, or by bond-tension failure in the vertical joints and tensile failure in the blocks along the line of the vertical joints. Two-core block could provide somewhat higher resistance to the latter type of cracking, according to Menzel [J.8].

J.5 □ A survey of cracking in two- to nine-story apartment houses

An extensive survey on cracking of apartment houses with external walls of masonry was carried out by the Johnson Engineering firm in cooperation with masonry unit manufacturers and builders in Sweden [J.14]. The investigation consisted of mapping cracks and studying the causes of cracking in 150 buildings. The buildings had different types of external plastering, ranging in thickness from $\frac{1}{10}$ to $\frac{1}{2}$ in. The masonry units were of the Swedish light weight cellular concrete type. Some of them were laid in lime mortar, some in lime–cement mortar, and some were laid without any mortar. The concrete slabs rested directly on the walls.

	Description	Type of crack	Number of cracks of type / All crack observations (%)
1a	Horizontal cracks at the topmost slab		12.1
1b	Horizontal cracks at intermediate floor slabs		9.9
2	Cracks in the foundation		13.3
3	Vertical cracks at the corner of the building		10.2
4	Horizontal cracks around the windows of the uppermost story of building		9.5
5	Cracks between roof and wall		8.8
6a	Cracks around staircases		8.3
6b	Cracks around balconies		5
7	Cracks in the mortar joints in the masonry		2.9
8a	Vertical cracks below window openings		9.9
8b	Horizontal cracks in the portion of the wall below the window		3.9
9	Horizontal cracks at the building corners		2.0
10	Vertical cracks at slab levels		4.2

Figure J.22 ☐ *Observed types of cracks from a survey on 150 buildings [J.14].*

The result of the survey is condensed in Fig. J.22 in which the different types of observed cracks with typical crack configurations are shown. The frequencies of the different types of cracks are also shown in Fig. J.22. ("Frequency" is expressed as the percent of the observations of the specific crack type related to all crack observations: one crack observation is noted for one particular type of crack on one building, but additional obser-

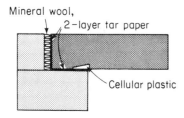

Mineral wool,
2-layer tar paper

Cellular plastic

Figure J.23 ☐ Suggested detailing at the topmost joint between floor slab and wall. The highly deformable insert permits rotation of the slab end and the tar paper permits sliding to take up differential shrinkage [J.14].

vations of the same type of crack on the same building are not counted.) The different types of observed cracks are discussed in the subsections which follow.

Some measures for crack prevention had been taken on the buildings. One type of preventive measure was to remove the usually rigid connection between the shrinking concrete slabs and the external wall by inserting tar paper between the slab and the wall. Another preventive measure was to tie the concrete slabs and the external wall together with concrete columns at the corners of the building. (Sometimes vertical ties were used.) From the survey, it was concluded that the method of freeing slab and wall from each other was superior. A complete interaction between the external wall and the rest of the building was evidently very hard to obtain.

Type 1a: Horizontal cracks at the topmost slab   This type of crack is due to differential movement in the concrete slabs and the external walls. Part of this differential movement comes from the high shrinkage of the concrete. The shrinkage in the concrete is about 0.5 per mil, but the light weight cellular concrete shrinkage is about 0.1 per mil. Another type of differential movement was discussed in Section J.2. The suggested remedy is to free the slab and the wall from each other completely, as shown in Fig. J.23. The cracking that will occur due to the sliding in the reentrant corner must be covered by repainting or by mouldings after the main portion of the shrinkage has taken place.

Type 1b: Horizontal cracks at intermediate floor slabs The causes for this type of crack are the same as for crack type 1a. This type of crack occurs mainly where the vertical force in the wall is low, for example, under large windows. To prevent this type of cracking, large windows should be avoided, especially close to building corners. Furthermore, the concrete floor slabs should have short spans and the slab concrete should exhibit low shrinkage. It is possible that tar paper inserts of the type shown in Fig. J.23 could be useful, but no practical experience has been gained as yet.

Type 2: Cracks in the foundation wall   This type of crack occurs mainly at reduced sections of the foundation walls and the causes are probably poor foundation, section changes, and stress concentrations, eventually in combination with temperature differences. This type of crack could probably be

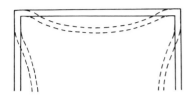

Figure J.24 □ Assumed mechanism for forming of vertical cracks at the corners of a building, type 3 in Fig. J.22 [J.14].

avoided by better survey of the building site and by proper design of the foundation, as well as use of low shrinking concrete.

Type 3:   Vertical cracks at the corner of the building   The mechanism for the occurrence of this type of crack is explained by Lantz [J.14], as shown in Fig. J.24. When the concrete slab shrinks, the external walls are relatively free to follow the slab between the corners, but at the corners the movement is prevented by inplane forces in the external walls. The result is a deflection of the external walls, shown by the dotted lines in Fig. J.24. High bending moments occur at the corners of the building, and, since the section of the external wall usually is weakened (by the bond pattern), these moments cause a vertical crack one unit away from the corner.

Type 4:   Horizontal cracks around the windows of the top story of the building   The main cause for this type of cracking is probably the moment in the external wall. See also Fig. J.2. This moment is caused by deflection of the concrete slab due to loading and uneven shrinkage. Close to the top of the building the normal force in the wall is low, and the load eccentricity is high; therefore, cracks will occur at a weak section of the wall. A short slab span should diminish the risk of this type of crack. The load eccentricity in the wall could also be diminished by a porous support part, as shown in Fig. J.23. This porous insert should be at least one inch thick; it should be made of a very soft material which takes practically no forces even for a relatively large compression.

Type 5:   Cracks between roof and wall   The cause of this type of crack is the differential movement between the roof structure and the wall. To prevent this type of crack, a free space of at least $\frac{1}{2}$ inch should be designed between the roof structure and the external wall.

The shrinkage of a wooden roof structure is more than 1 per mil parallel with the wood fibers; the difference in relative humidity during the seasons of the year can result in a movement of about 0.4 per mil in the wood. Perpendicular to the fiber direction, the movements dependent on humidity are 10 to 20 times greater. Regardless of the material used in the roof, the thermal movements of the roof structures are much higher than the movement of the external walls. This is an additional reason for permitting the roof structure to move independently of crack sensitive external walls.

Type 6a: Cracks around staircases   This type of crack appears at the lower part of the building around the entrances. These cracks are probably caused by the section changes which occur in the wall and the slabs at the entrances of a building. Both the foundation structure and the internal wall as well as the slabs are often weakened by the entrance opening in the structure. A low shrinkage concrete and a suitable reinforcement in the foundation could help prevent this type of cracking.

Type 6b: Cracks around balconies   This type of cracking has the same causes as type 6a. In addition, the movements of protruding concrete slabs which are unshielded against temperature variations, and which therefore have yearly and daily thermal movements outside the building, often result in such cracks. Protruding balcony slabs should be insulated from the rest of the building by, for example, some kind of bituminous material, or by rubber, cork, etc.

Type 7: Cracks in the mortar joints   The cause for this type of cracking is mainly the differential movement in the plaster layer and the masonry itself. The movement in the plaster comes from temperature variations and variations in the moisture content (due to rain or difference in relative humidity). The heat radiated by the sun during the day can quickly warm up the plaster layer and cause expansion, which later on extends inward in the wall until the large part of the wall is heated and expanded; then, when the radiation reverses at night, the plaster quickly cools off. Sometimes the plaster is unable to withstand the tensile stress that occurs during contraction. The effect is less pronounced for thicker plastering. The plaster should be from $\frac{3}{8}$ to 1 inch thick, built up of three layers, according to Lantz [J.14].

Type 8a: Vertical cracks below window openings   One of the causes for this type of crack is probably a movement in the foundations. The difference between the very high load in the parts of the wall on the side of the windows and the absence of load in the part below the window could also cause excessive stresses in the cracking part of the wall. Furthermore, radiators located below windows causes rapid drying of this part of the wall, facilitating cracking. A better foundation, eventually in combination with the reinforcement in one of the joints immediately below the window, could prevent this type of crack.

Type 8b: Horizontal cracks in the portion of the wall below the window   These cracks follow the horizontal joints below the window. The stress differences between the portion of the wall to the side of the window and the portion of the wall below the window, and rapid drying of the portion of the wall below the window, could be the causes of this type of cracking. Thick plastering as well as slow drying of this portion would help prevent this type of crack.

Type 9:   Horizontal cracks at the building corners   The cause for this type of cracking is the vertical lift of the slab corners, well-known from the theory of slabs. If the slab is freed from the wall by a bituminous layer, then the slab can lift freely without cracking the wall. The inside crack must be covered by repainting or mouldings.

Type 10:   Vertical cracks at slab levels   This type of cracking occurs when the thin masonry units outside the slab edge are poorly bedded in the mortar joints and heavily loaded from above. The cracks can also occur when the thin masonry units have been disturbed during the pouring of the concrete slabs.

Although all types of cracks described above have occurred in light weight cellular concrete masonry walls, most of the phenomena could occur in any type of masonry to a greater or lesser degree. The lessons learned from this survey should, therefore, help in obtaining a good design for any type of external masonry wall.

References for Chapter J

J.1 □ Sahlin, Sven: "Utvändiga väggsprickor vid vindsbjälklag." (External Cracks in Walls at the Top Floor Slab in a Building.) Väg-och vattenbyggaren No. 3, Stockholm, 1964.

J.2 □ Nevander, Lars Erik: "Tekniska Egenskaper hos Isolerade Hålmurar av Tegel." (Technical Properties of Clay Brick Cavity Walls.) Institutionen för Byggnadsteknik Kungl., Tekniska Högskolan, Stockholm, 1961.

J.3 □ *Lättbetonghandboken 1965.* (Handbook of Light Weight Cellular Concrete.) Stockholm, 1965.

J.4 □ ACI Committee 716: "Physical Properties of High-Pressure Steam-Cured Block." Journal of the American Concrete Institute, Proceedings, Vol. 49, pp. 745–756, April, 1953.

J.5 □ Kuenning, William H., and Carlson, C. C.: "Effects of Variations in Curing and Drying on the Physical Properties of Concrete Masonry Units." Development Department Bulletin D13, Portland Cement Association, December, 1956.

J.6 □ Shiedeler, J. J.: "Carbonation Shrinkage of Concrete Masonry Units." PCA Bulletin D69, 1963.

J.7 □ Verbeck, G.: "Carbonation of Hydrated Portland Cement." Papers on Cement and Concrete, American Society for Testing Materials, Special Technical Publication 205, 1958.

J.8 □ Menzel, Carl A.: "General Considerations of Cracking in Concrete Masonry Walls and Means for Minimizing It." PCA Development Department Bulletin D20, September, 1958.

J.9 □ Hedstrom, R., Litvin, A., and Hanson, J.: "Influence of Mortar and Block Properties on Shrinkage Cracking of Masonry Walls." Journal of the PCA Research and Development Laboratories, Vol. 10, No. 1, January, 1968.

J.10 □ Dührkop, Saretok, Sneck, and Svendsen: "Bruk-Murning-Putsning." (Mortar-Masonry-Plastering.) National Swedish Council for Building Research, Stockholm, 1966.

J.11 □ Nylander, H., and Ericsson, E.: "Inverkan av Deformationer i Väggar på Bjälklagslaster och Bjälklagsdeformationer vid Flervåningshus." (Effects of Wall Deformations on Floor Slab Loads and Floor Slab Deformations in Multi-Story Houses.) Nordisk Betong, Vol. I, No. 4, p. 292, 1957.

J.12 □ Tegelindustriens Centralkontor AB.: "Tekniska data för tegel och Tegelkonstruktioner." Stockholm, 1965.

J.13 □ Nevander, L. E.: "Provningar av Tegelmurverk." (Tests on Brick Masonry Walls.) Tegel, Nr. 5, Stockholm, 1964.

J.14 □ Lantz, Håkan: "Hur Undvika Sprickbildning i Ytterväggar av Lättbetong." (How to Avoid Cracking in External Walls of Light Weight Cellular Concrete.) Byggnadsindustrin No. 14.67 (or Tekn. Dr. Arne Johnson Ingenjörsbyrå. Tekniska meddelanden Nr. 20), Stockholm, 1967.

K ☐ Reinforced masonry

K.1 ☐ Introductory remarks

High tensile stresses are induced in a structure under some loading conditions such as bending of beams. Since masonry has a fairly low tensile strength and low ductility, reinforcement is needed in high tension areas of a masonry structure.

The behavior of reinforced masonry is quite similar to that of reinforced concrete. However, the orthogonal anisotropic nature of masonry causes some differences under certain types of loading. Overreinforced masonry beams are more brittle than overreinforced concrete beams, and the relative magnitudes of shear tensile and compressive strength for masonry are somewhat different from those for reinforced concrete. In the following sections the behavior of masonry will be discussed, under the assumption that the reader is familiar with fundamentals of reinforced concrete. The treatment will be relatively brief since information about the basic principles is readily available in textbooks on reinforced concrete.

The moment-carrying capacity of reinforced masonry beams is dealt with in Section K.2, the shear strength of reinforced masonry beams in Section K.3, and the bond strength of reinforcing bars in Section K.4. The selection of mortar is discussed in Section K.5. Section K.6 is devoted to reinforced masonry columns, and Section K.7 to ductility of reinforced masonry. Structural details are discussed in Chapters L and M.

K.2 ☐ Moment capacity of reinforced masonry beams

Withey [K.1] described tests on brick masonry beams with steel percentages in the range 0.56 to 2.31 and different types of bricks with properties as shown in Table K.1. The strength of the compressive zone in the masonry beams was tested on masonry prisms with the bricks in the same position as in the beams. The compressive stresses acted parallel to the longest axis of the bricks. The results of the 24 beams tested are plotted in Fig. K.1 and compared with the well-known Whitney-type formula for the ultimate moment in a reinforced concrete beam:

$$m = q(1 - 0.59q) \lesssim 0.4 \tag{K.1}$$

where

$$m = \frac{M}{bd^2 f_m'} \quad \text{and} \quad q = \frac{pf_y}{f_m'}$$

b = width of beam; d = effective depth of beam; M = ultimate moment; f_m' = strength of masonry in compression; p = steel area divided by masonry area (width times effective depth); f_y = steel yield stress. [Because of brittleness observed in some tests and limited experimental experience with overreinforced masonry beams, the calculated ultimate moment m is limited to about 0.4 in Eq. (K.1)]. The calculated and the observed ultimate moments (Fig. K.1) are in good agreement except for beams C7 and C8, both of which had poor bonding between steel and bricks due to grooved joints. This kind of failure is not uncommon in tests of reinforced beams; it strongly indicates that the main concerns in designing reinforced masonry beams should be the shear and bond strength of the masonry and the quality of work-

Figure K.1 ☐ Ultimate moment of beams tested by Withey [K.1], comparison with Eq. (K.I). See also Table K.1.

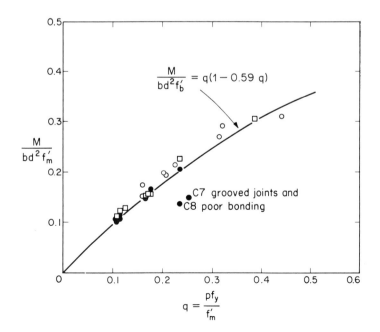

Notations in Chapter K

τ = shear stress

V = shear force in a reinforced masonry beam

b = width of beam

d = effective depth of beam

jd = distance between the resulting force in the compression zone of a beam and the bars (resulting tensile force)

a = shear span (equals distance between support and nearest load on the span of a beam)

p = ratio between area of steel and effective area $b \cdot d$ of beam

f'_y = yield point stress of steel

f'_m = crushing strength of masonry

M = bending moment on beam

---

manship in the brick laying. A calculation according to the classical elastic theory of reinforced concrete beams will also give reasonable results for masonry, since at least the higher strength masonries are fairly linearly elastic and sometimes even brittle in the compression zone of beams.

Tests by Cox and Ennenga [K.2] on laterally loaded reinforced masonry walls show that the bending theory for reinforced concrete applies also in this case. The reported ultimate moments of three walls correspond to a tensile steel stress of 105,000 psi at failure, which indicates tensile failure of underreinforced beams (about 0.031% reinforcement). It has also been shown in

Table K.1 □ *Strength of bricks and masonry prisms in Withey's tests [K.1]. See Figure K.1. Mortar mix: 3 : 1 : 12 by weight. The compressive strength of mortar cylinders averaged 184 kg/cm² (2833 psi). The tensile strength of mortar (28 days) averaged 24.8 kg/cm² (354 psi). Some beams had shear reinforcement.*

Type of bricks	Position of bricks	f'_{brick} (psi)	Modulus of rupture (psi)	Absorption after 5 min. immersion in water—%	Prisms endwise $f'_{masonry}$ (psi)	Mark in Figure K.1
Chicago	Flat	2,586	971			
	Edge	2,390		7.7		
	End	5,550			2,690	●
Waupaca	Flat	12,500	2,250			
	Edge	5,860		2.2		
	End	5,900			2,870	□
Streator	Flat	9,195	1,110			
	Edge	4,034		2.9		
	End	3,970			2,090	○

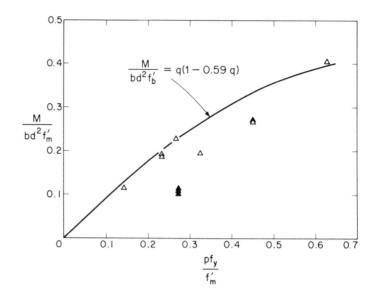

Figure K.2 ☐ Ultimate moment of beams tested by Granholm [K.3]; Δ = bricks with compressive strength of 158 kg/cm² (2250 psi) and mortar compressive strength 312 kg/cm² (4410 psi) giving a masonry with the compressive strength 82 kg/cm² (1170 psi); ▲ = mortar strength of 118 kg/cm² (1680 psi).

several other tests (see, for example, Granholm [K.3], Fig. K.2) that the ultimate moment can be predicted reasonably well with the aid of Eq. (K.1) except when the shear strength and bond strength of the beam are insufficient, as for the points under the curve in Fig. K.2. These latter factors are dealt with in the following section.

K.3 ☐ Shear strength of masonry beams

The shear strength very often determines the ultimate load-carrying capacity of a masonry beam, and thus shear strength is often the critical strength property of masonry.

Figure K.3 ☐ Ultimate moment of beams tested by Parsons, Stang, and McBurney [K.4], comparison with Eq. (K.1). See also Table K.2. Beams below the curve failed in shear. See Fig. K.4 for nominal shear stresses.

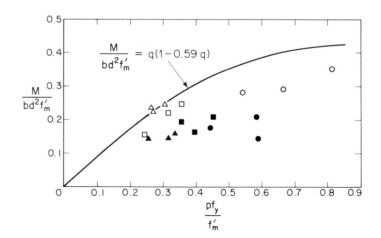

Table K.2 ☐ *Tests by Parsons, Stang, and McBurney [K.4]. See Figures K.3 and K.4. Mortar: 1 part cement, 3 parts sand plus hydrated lime, 15% of cement volume. Flow: 108%. Compressive strength: 165 kg/cm² (2340 psi) dry storage, 263 kg/cm² (3740 psi) damp storage.*

Type of bricks	Position of bricks	f'_{brick} (psi)	Modulus of rupture (psi)	Absorption after 5 hrs. immersion in water—%	Prisms $f'_{masonry}$ (psi)	Mark in Figure K.3
Chicago	Flat	3910	1530	8.8	1021	△
	Edge	4280			1856	☐
	Combined				2025	○
Philadelphia	Flat	4510	650	11.1	1202	▲
	Edge	5240			1688	■
	Combined				1389	●

The shear strength of masonry beams was tested by Parsons, Stang, and McBurney [K.4] on beams made of Chicago- and Philadelphia-type bricks. The bonding arrangement was varied so that one type of beam had the bed joint parallel to the span, one type of beam had the bed joint perpendicular to the span, and one type of beam had the upper half parallel to the span and the lower half perpendicular to the span. The brick properties as well as the mortar properties are listed in Table K.2. To obtain values for the strength of the compressive zone in the beams, masonry piers were tested with the same bonding arrangements and the same stress direction as in the beams. The resulting pier strengths are listed in Table K.2. To check the type of failure, all the test results are plotted in Fig. K.3 are compared with the ultimate moment curve obtained from Eq. (K.1). Figure K.3 indicates that all the beams of type CA failed in flexure. The other beams did not reach their ultimate bending capacity because of shear failures at lower loads. The underreinforced beams of type CB were not to far from flexural failure, but the overreinforced beams of type CC were short of flexural failure by a substantial amount.

In Fig. K.4 the nominal shear stress $V/(j\,db)$ is plotted as a function of the shear span ratio a/jd for the beams whose moment capacities are plotted in Fig. K.3. The test results (Fig. K.4) indicates that the shear stresses were inversely proportional to the shear span ratio, a phenomenon also observed for reinforced concrete. (The highest nominal shear stresses were obtained for the beam that probably failed in bending, which indicates that the shear strength was sufficiently high to produce failure in bending.) The lowest nominal shear strength was obtained for

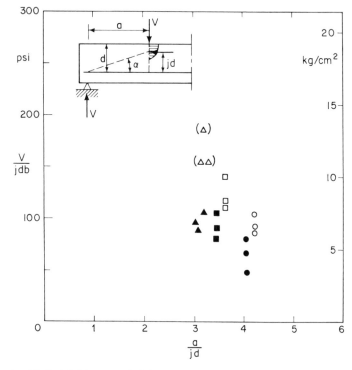

Figure K.4 ☐ Nominal shear strength of masonry beams tested by Parsons, Stang, and McBurney [K.4] as a function of the shear span ratio a/jd. Beams within parentheses failed in bending. See also Fig. K.3 and Table K.2.

the Philadelphia-type brick. Furthermore, the lowest strength was obtained for the beams with the bed joints perpendicular to the span. The lowest value was 48 psi, which is about 4% of the compressive strength of the corresponding masonry piers. The percentage for Chicago-type brick was about 8.5.

The lowest shear strength for the beams having the bed joints parellel to the beam span was 88 psi for Philadelphia-type masonry. This was 5.3% of the compressive strength of masonry piers. The lowest value for Chicago-type bricks was 154 psi (for beams failing in bending) corresponding to over 15% of the compressive strength of masonry piers. The better performance in shear of the Chicago-type bricks is probably due to the lower suction of this type of brick.

The range of shear span ratio variation in Fig. K.4 is very small and several test series have to be taken into account to cover a large range, although at the same time other factors that are not the same from one test series to another, impair a comparison. Hilsdorf [K.5] compiled test data reported by several authors (Fig. K.5). In Fig. K.5 the results from tests by Zelger [K.6] have also been added (see also Table K.3). It should be noted that in Fig. K.5 the shear span is defined as the distance *a* from the support to the load, divided by the effective depth of the beam from the steel bars to the upper edge of the beam. Although the scatter of the tests results is considerable, the general influence of the shear span ratio is clearly seen and it has the same general trend as for reinforced concrete beams. The increase in shear strength for short shear span ratios is partly due to the increase in normal stresses perpendicular to the bed joints as the angle α in Fig. K.4 increases. The increased vertical stress component increases the "friction" capacity of the joint.

On the basis of these tests [K.6] and the tests [G.3] described in Chapter G, Zelger proposed the following equation for calculation of the nominal shear stresses in a masonry beam

$$\tau = \frac{V}{j\,db}\left(1 - \frac{\mu}{\lambda}\right) \tag{K.2}$$

where μ is the friction coefficient ≈ 0.5 and λ = shear span ratio $= a/z = a/jd$ [0.8 = minimum value to be inserted in Eq. (K.2)] The shear strength of masonry beams for shear span ratios

Figure K.5 □ Nominal shear strength of masonry beams as a function of the shear span ratio a/d (from Hilsdorf [K.5] and Zelger [K.6]). Compare Table K.3.

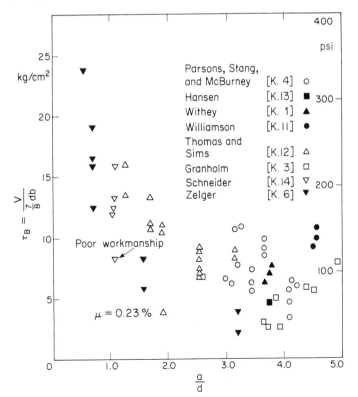

Table K.3 ☐ *Tests by Zelger* [K.6].
See Figure K.6.

Type of bricks	Position of bricks	f'_{brick} (kg/cm²)	Mortar f' (kg/cm²)
Cored	Flat	206	20–30
	End	59	

over three were probably on the order of 10% of the compressive strength of the masonry parallel to the beam span. However, any data for the masonry strengths parallel to the bed joints were not reported by Zelger but are estimated on the basis of the brick strength and the mortar strength. The corresponding value obtained from Granholm tests was about 7.5%.

It seems reasonable on the basis of the reported data to assume that the nominal shear strength of masonry beams is 40 psi or more (or 4 to 10% of the compressive strength) at high shear span ratios and about double this figure for high strength mortar in combination with cored high strength brick. For shorter shear spans, the increase in shear strength can be estimated with the aid of Eq. (K.2). It should, however, be emphasized that the shear failure phenomenon of reinforced masonry beams is still not well understood. As a guide for the estimate of the shear strength of masonry beams, the following factors should also be considered.

The absorption (complete immersion) of the bricks influences the shear strength as shown in Fig. K.6, compiled by Hilsdorf [K.5]. The maximum shear strength occurs for 10% absorption; a "dusty" or spalling brick surface can reduce the strength by about 50%. All tests in Fig. K.6 except those by Withey were run on solid bricks. Too few tests are known to permit comparisons for cored bricks.

The detrimental effect of high suction seems to be somewhat counteracted by a wet mortar, so that mortars with high water to cement ratios give higher shear strengths, even through the compressive strength of the mortar decreases with increasing water to cement ratio. An increased *cement* content in the cementitious material in the mortar increases the shear strength considerably, as can be seen in Fig. K.7, compiled from test data by Granholm [K.3]. The mortars had probably about equal flows to permit easy laying, but varying water to cement ratios, and therefore the illustration should be used with some caution.

With increasing amount of tensile *reinforcement* the shear strength of the beam also increases (see Fig. K.8 after Hilsdorf).

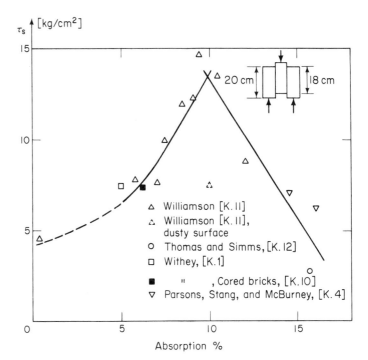

Figure K.6 ☐ Nominal shear strength of masonry (brick triplets) as a function of absorption (by weight), Hilsdorf [K.5].

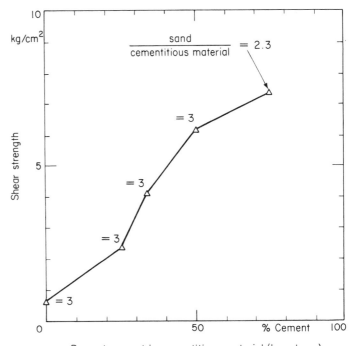

Figure K.7 ☐ Nominal shear strength of masonry (brick triplets) as a function of percent cement in the cementitious material. The ratio of sand to cementitious material varied in the tests, Granholm [K.3].

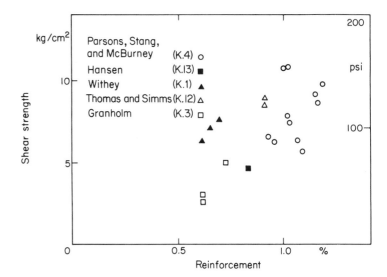

Figure K.8 □ The nominal shear strength of masonry beams as a function of the percentage reinforcing, Hilsdorf [K.5]. Shear span ratios of 3 to 4.

This effect, also known from reinforced concrete beams, comes from the increase in compression zone with increasing percentage of reinforcement, with decrease in vertical crack lengths; possibly it is also due to some extent to dowel effect of the bars. The test data for reinforced masonry beams are too few to be conclusive, and Fig. K.8 should be read with some caution for this reason and also because the data were compiled from several different test series which show differences among other variables.

The shear strength of reinforced masonry beams is to some extent correlated to the *shear strength of brick triplets*, as shown in Fig. K.9, compiled by Hilsdorf [K.5].

Reinforced masonry beams can be protected against shear failure by stirrups when the shear strength of the plain masonry is insufficient. The practical problems are considerable, since the stirrups cannot be placed freely but only in vertical joints or holes in the masonry units. By careful planning, the method can be used with reasonable ease. As a tentative recommendation we may suggest that the shear reinforcement be designed to take the whole shear force in sections where it is needed.

K.4 □ Bond between reinforcing bars and mortar

Reinforced masonry beams can also fail in bond between steel and mortar. Experimental data on beams (except for observa-

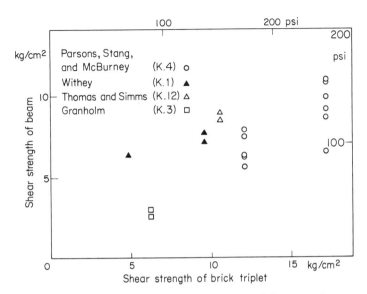

Figure K.9 □ *The nominal shear strength of masonry beams as a function of shear strengths observed on brick triplets of the same materials, Hilsdorf [K.5].*

tions in test series to study bending or shear failures, as for example in Fig. K.1) are few, but some pullout tests are reported in the literature. Plummer and Reardon [K.7] report data obtained by Osborn on $\frac{3}{8}$-in. round steel bars embedded 3 to 3.7 in. in 1 : 3 grout in cavity walls. The bond stress at 0.0001 in. slip was about 50 kg/cm² (700 psi) and at ultimate load 120 kg/cm² (1700 psi), which was about 120% of the yield stress of the bars. (These bars had a short embedment length, and they were deformed, so they were probably well bonded to the grout. These factors should result in high bond strength.)

Gallagher [K.8] reports tests on pullout specimens with two bricks plus a mortar joint in which the steel bar was embedded along a length of 18 in. The ultimate stresses ranged from 14 to 40 kg/cm² (200 to 550 psi). The lowest value was for a $\frac{3}{8}$-in. bar and $\frac{5}{8}$-in. joint with a low strength mortar, and the highest values were found for $\frac{3}{8}$-in. bars embedded in $\frac{1}{2}$-in. high strength mortar joints. Tests on $\frac{1}{2}$-in. square bars were performed by Parsons, Stang, and McBurney [K.4]. The embedded length was shorter in this case, about 8 in. The reported strengths are higher than in Gallagher's tests. Granholm [K.3] ran pullout tests on 1-cm round bars and varied the cement to lime ratio in the mortar and found bond strengths 3 to 42 kg/cm² (40 to 600 psi).

The tests reported by these three authors have been compiled in Fig. K.10. It is difficult to make direct comparison because

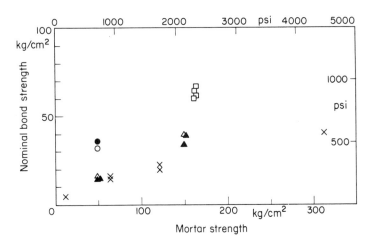

Figure K.10 ☐ Relationship between mortar compressive strength and pullout bond strength according to Granholm [K.3], Gallagher [K.8], and Withey [K.1].

Source	Embedment length (cm)	Bar diameter (cm)	Mortar bed thickness (cm)	Mark
Granholm [K.3]	12.5	1.0		×
Gallagher [K.8]	45.5	0.63	1.25	○
"	45.5	0.63	1.59	●
"	45.5	0.95	1.25	△
"	45.5	0.95	1.59	▲
Withey [K.1]	20.3	1.25 square		☐

the mortar strengths were not tested in the same manner by the different authors. For example, Gallagher tested the mortar strength in two different manners and obtained mortar strength about twice as high for wet curing as for dry curing (shown in Fig. K.10). From the figure the following conclusions can be made. The bond strength is about 15% of the mortar strength for 1-cm round bars. Higher values are obtained for thinner bars, for square bars, and for deformed bars in grout.

K.5 ☐ Choice of mortar

The best mortar for reinforced masonry must be chosen with several factors in mind. The mortar should contain much cement to produce a high strength, but enough lime to ensure good workability. The total amount of cementitious material should be limited to keep the costs down; the sand should be well graded to minimize the shrinkage. One should strive for a high flow mortar with high water retentivity. Mix proportions of cement to lime to sand by volume of $1 : 0.3 : 3$ (Hilsdorf), $1 : 0.25 : 2.25$ (SCPI), $1 : 0.5 : 3$ (SCPI), $1 : 0.3 : 5$ (Withey), and $1 : 1 : 3$ (Granholm) have been recommended.

K.6 ☐ Reinforced brick columns

Tests by Lyse [K.9] showed that the strength of a reinforced brick masonry column may be computed from the formula

$$P_{ult} = A(k \cdot f_b' + p_s')$$ (K.3)

where P_{ult} is total strength of column; A is total area of column; k is "efficiency factor," i.e., ratio between strength of masonry and strength of bricks; p is ratio between area of longitudinal steel and of column; f_b' is compressive strength of bricks; and f_s is yield point stress of steel.

Ties in every fourth bed joint were considered [K.8] to be sufficient to develop the yield stress of the longitudinal steel. Large reinforcing bars, however, added only a portion of their yield stress to the strength of the column, probably due to difficulties with the embedding of the longitudinal reinforcement, which therefore had poor bonding and did not interact with the masonry. Withey [K.10] derives a formula similar to Eq. (K.3) on the basis of his own tests. Withey's formula, however, also takes into account the increase in column strength caused by lateral reinforcement in the form of hoops surrounding the longitudinal bars.

K.7 ☐ Ductility of reinforced masonry

The load–deflection relationship for reinforced masonry beams is similar to that for reinforced concrete beams. See Fig. K.11, from Granholm [K.3]. In the precracking region the relationship follows the theoretical one (Stadium I) calculated according to the theory of elasticity, assuming the beam to be fully elastic and the modulus of elasticity of steel to be 35 times the modulus of elasticity of masonry. After cracking, the observed relationship gradually approaches the calculated (Stadium II). The maximum load is reached by considerable inelastic deformation,

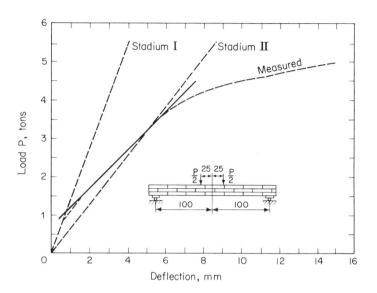

Figure K.11 □ Load deflection diagram for beam No. 3 tested by Granholm [K.3]. The reinforcement consisted of seven 10-mm bars. The theoretical curves, Stadium I referring to the uncracked stage and Stadium II referring to the cracked stage, are calculated with the assumption that the modulus of elasticity of the steel is 35 times the modulus of elasticity of the masonry.

as can be seen from the nearly horizontal portion of the curve in Fig. K.11. A nearly linear relationship was observed for a considerable portion of the measured load-deflection curve as indicated by the solid tangent line in the figure.

The columns tested by Withey [K.10] also underwent inelastic deformations before failure. The ultimate strain was, in one instance, for a column with 0.75% lateral reinforcement and 1.93% longitudinal reinforcement, about 0.004 at a stress of $140 \, \text{kg/cm}^2$ (2000 psi) and in one column without reinforcement about 0.0008 at a stress of $50 \, \text{kg/cm}^2$ (700 psi).

References for Chapter K

K.1 □ Withey, M. O.: "Tests on Brick Masonry Beams." ASTM, Proceedings of the Thirty-Sixth Annual Meeting, Vol. 33, Technical Papers, Part II, pp. 651–669, 1933.

K.2 □ Cox, F. W., and Ennenga, J. L.: "Transverse Strength of Concrete Block Walls." Journal of the American Concrete Institute, Vol. 29, No. 11, May, 1958.

K.3 □ Granholm, H.: "Armerade Tegelkonstruktioner." Transactions of Chalmers University of Technology, Nr. 16, Gothenburg, Sweden, 1943.

K.4 □ Parsons, D. E., Stang, A. H., and McBurney, J. W.: "Shear Tests of Reinforced Brick Masonry Beams." U. S. Department of Commerce, Bureau of Standards, Journal of Research, Research Paper 504, Vol. 9, pp. 749–768, 1932.

K.5 □ Hilsdorf, Hubert: "Bewehrtes Ziegelmauerwerk." Materialprüfungsamt für das Bauwesen der Technischen Hochschule München, Bericht Nr. 33, 1962.

K.6 □ Zelger: "Neue Wege bei der Bemessung von Ziegelsturzen." Die Ziegelindustrie, Heft 4, 1964.

K.7 □ Plummer, Harry, and Reardon, Leslie: "Principles of Brick Engineering." Structural Clay Products Institute, Washington, D. C., 1943.

K.8 □ Gallagher, E. F.: "Bond Between Reinforcing Steel and Brick Masonry." Brick and Clay Record, Vol, V, pp. 86–87 and 92–93, March, 1935.

K.9 □ Lyse, Inge: "Tests of Reinforced Brick Columns." Journal of the American Ceramic Society, Vol. 16, pp. 584–597, 1933.

K.10 □ Withey, M. O.: "Tests on Reinforced Brick Masonry Columns." ASTM, Proceedings of the Thirty-Seventh Annual Meeting, Vol. 34, Technical Papers, Part II, pp. 387–405, 1934.

K.11 □ Williamson, H. D.: "Reinforced Brickwork." Rensselaer Polytechnic Institute, Engineering and Science Series No. 46, Troy, New York, 1934.

K.12 □ Thomas, F. G., and Simms, L. G.: "The Strength of Some Reinforced Brick Masonry Beams in Bending and in Shear." The Structural Engineer, p. 330, July, 1939.

K.13 □ Hansen, James A.: "Developments in Reinforced Brick Masonry." Proceedings of the American Society of Civil Engineers, p. 407, March, 1933.

K.14 □ Schneider, R. R.: "Grouted Brick Masonry—Report of Tests." Southwest Builder and Contractor, December, 1951.

L □ Earth pressure
on foundation walls

L.1 □ Introductory remarks

The more durable types of masonry are often used as foundation
walls, especially for small houses. Two major problems are involved in the design: the calculation of the earth pressure against
the wall, and the resistance of the wall against this pressure. In
practice, the future overload on the earth surface, excavations,
water flooding, etc., are not too well-known. Therefore, normally only an approximate calculation of the earth pressure on
the safe side can be made, since the basic assumptions for the
earth pressure theories are also somewhat uncertain. This uncertainty is also related to the uncertainty of the boundary conditions and failure mechanisms for the foundation wall itself.
These factors are discussed below.

L.2 □ Earth pressure against foundation walls

The backfill behind a foundation wall usually consists of some
kind of friction material such as sand and gravel, sometimes of
not too well-defined composition. The fill must at least be per-
vious enough to drain rain water from behind the foundation
wall. At the ultimate stage, before a foundation wall completely
breaks in, both the wall and the soil have deformed to some
extent. Then the soil mass is said to be in an active stage of
stress and in plastic equilibrium.
   The active earth pressure, from an ideal friction material, can
be calculated according to Rankin's classic formula:

$$p = \tfrac{1}{2}\gamma H^2 \tan^2(45° - \tfrac{1}{2}\phi) = \tfrac{1}{2}\gamma H^2 K_A \qquad (L.1)$$

where p is the unit pressure against the wall, γ is the density of
the soil, H is the depth under the soil surface, ϕ is the internal

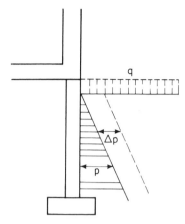

Figure L.1 □ Earth pressure against a foundation wall.

angle of friction of the soil, and K_A is $\tan^2 (45° - \phi/2)$. According to Eq. (L.1), the earth pressure will increase linearly from zero at the surface to a maximum value at the bottom end of the foundation wall (Fig. L.1). A detailed discussion of Eq. (L.1) can be found in textbooks on soil mechanics, e.g., Terzaghi and Peck [L.1]. See also books on earth pressure and retaining walls, Huntington [L.2].

If the backfill carries a superimposed load q, the additional pressure against the foundation wall is

$$\Delta p = q \cdot K_A \tag{L.2}$$

to be added to the pressure calculated according to Eq. (L.1) (Fig. L.1).

In practical applications some questions can be raised as to whether the pressure values calculated from Eq. (L.1) are valid. For example, the foundation wall can be supported at its top by a building. This drastically reduces the movement of the top of the wall. Then it is questionable for stiff walls whether the deformations are large enough to justify the assumption of active soil pressure, and the larger passive state of stress should be considered. On the other hand, the firm support of the top and bottom of the wall could permit some kind of arching action to take place in the soil body itself as well as in the wall. Such phenomena would reduce the earth pressure against the wall. Similarly, horizontal arching action could take place, at least over short walls. The excavation for the basement of a house is usually only a few feet bigger than the basement itself, and the sand backfill extends only a few feet outside the foundation walls. From there on, the soil is often of unknown quality at the time of design. This causes an uncertainty in the calculations. The uncertainties mentioned justify the presentation of test results from tests on actual foundation walls for town houses.

Notations in Chapter L

H = depth under soil surface

$K_A = \tan^2 (45° - \phi/2)$

p = unit pressure against a foundation wall

q = unit superimposed load on soil surface

ϕ = internal angle of friction of soil

γ^2 = density of soil

Tests by Broms and Rehnman [L.3] showed that the load caused by backfill of sand on foundation walls for some town houses was approximately triangularly distributed. For the loosely filled sand, the measured earth pressure was 1 to 1.5 t/m² (1.5 to 2 psi) at a depth of 2.2 m (7 ft 3 in.) below the surface. The measured loads were in reasonable agreement with the classical earth pressure theory calculated for an internal friction angle of 30° and a density of 1.7 kg/dm³ (106 lb/ft³) for the sand. The tests were performed on foundation walls with a length of 7 m (23 ft) and a height of 2.5 m (8 ft 3 in.). The maximum deflection was about 4 mm. The wall was supported on three sides only. The deflections for walls which were supported by an intermediate cross wall were much smaller (less than 1 mm). After compaction by an 8-ton tractor, the earth pressure had increased

Figure L.2 □ *Earth pressure against a foundation wall and movement of the wall. The upper diagram shows the pressure in t/m² = Mp/m² (1t/m² = 205 lb/ft²). The lower diagrams show the deflection of the wall (in mm) as a function of the depth. Diagrams ④ and ⑥ refer to the ends of a 4m (13 ft) long wall and the diagram ⑤ to the central line of the wall. The solid lines refer to the readings before compaction and the dotted lines, to the readings after compaction of the sand backfill Broms and Rehnman [L.3].*

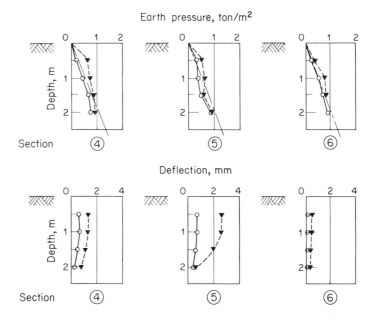

Earth pressure, ton/m²

Deflection, mm

closest to the surface. The increase in pressure was between 15 and 100%. At the same time, the porosity of the material decreased. Examples of the measured data are shown in Fig. L.2. In this illustration the theoretically calculated earth pressure is also plotted. The maximum span of the wall was in this case 4 m (13 ft). In Fig. L.3 the corresponding results for a 7 m (23 ft) wall is shown. In Fig. L.3 it can be seen that the deflections are exceedingly large after compaction (dotted lines), and this is an

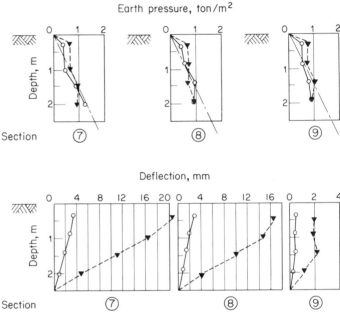

Figure L.3 ☐ (Above) Earth pressure against a foundation wall and movement of the wall. The upper diagrams show the soil pressure against the wall and the lower diagrams show the deflections of the walls. The wall was 7 m (23 ft) long. Section ⑦ refers to the center line of the wall: diagram ⑧, to the quarter points: and ⑨, to the end of the wall. The diagrams are according to Broms and Rehnman [L.3].

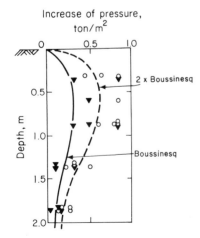

Figure L.4 ☐ (Left) The diagram shows the increase in earth pressure against a foundation wall as a function of the depth below the surface. The superimposed load consisted of an 8-t tractor placed 0.5 m (1 ft 4 in.) from the outside of the wall. The triangles refer to a compacted sand fill and the open circles refer to a loose fill. Computed values according to Boussinesque theory are shown by a solid line in the figure and two times the calculated value are shown by a dotted line.

indication that the wall had reached its ultimate strength in spite of the fact that the load was of the same magnitude as for the shorter wall in Fig. L.2. The compaction by the tractor increased the deformation gradually for each passage but the permanent load did not increase correspondingly.

In Fig. L.4 the increment in load as a function of the depth is shown for the first loading by the tractor, and also for a loading by the tractor after the sand had been compacted. (The tractor was placed at the section where the readings were taken and 0.5 m from the external face of the foundation wall.)

According to the authors [L.3], it seems reasonable to design a foundation wall for a lateral liquid pressure which is 40% of the weight of the soil, provided that the ground water table is below the lower edge of the foundation wall and that the sand fill is not compacted. Additional load from the superimposed loads, as for example from a tractor, can be calculated according to Boussinesq's theory (see textbooks on soil mechanics [L.1] and [L.2]) for the compacted fill, using twice the calculated values for the loose fill. These values are approximate, as can be seen from Fig. L.4.

A foundation wall should also be designed so that it can deflect about 0.2% of the wall height to permit the active earth pressure to develop. For highly compacted material in the fill, the deflection might be much less, 0.05% of the wall height. These figures apply to friction material, such as sand only. For less flexible walls, a higher pressure should be considered until more experimental evidence is gathered.

For calculation of the bending moment in the foundation wall, the reader is referred to textbooks on slabs. For one type of boundary condition the calculated bending moments are shown in Fig. L.5, prepared by NCMA [L.6]. In this case, the wall is hinged at the top and bottom and fixed at both ends. The moments can be calculated with the aid of the diagram shown in the figure. The obtained moment distribution and the stresses can then be judged against the test results reported in Chapter H, as well as the permissible stresses for bending given in the applicable building code.

A typical detailing of the bottom end of the basement wall is shown in Fig. L.6 [L.7]. In dry environments the concrete base and the membrane insulation can be omitted. If the floor cannot be assumed to support the wall laterally, a reinforcing bar should be inserted into the footing and extended into the wall.

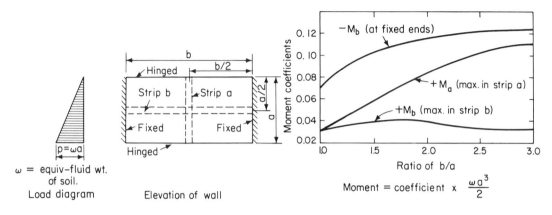

$$\omega = \text{equiv-fluid wt. of soil.}$$
Load diagram   Elevation of wall

Moment = coefficient x $\dfrac{\omega a^3}{2}$

Figure L.5 ☐ (Above) Moment coefficients for the middle strips of a foundation wall. The values refer to a case with fixed ends and simple supports at the top and bottom, according to the National Concrete Masonry Association [L.6].

Figure L.6 ☐ (Below) Typical detail at the bottom end of a wall of concrete block, Portland Cement Association [L.7].

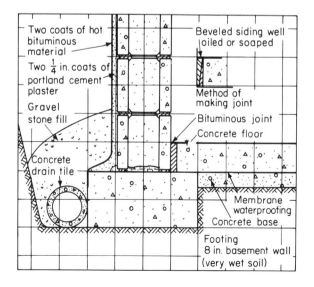

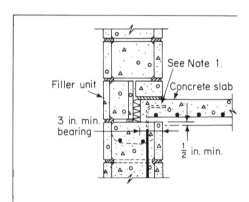

Notes:
1. Extend and embed every third or more vertical bars 6 in. into floor slab so as to provide anchorage at intervals not exceeding 48 in. Where the spacing of the vertical bars equals or exceeds 48 in., anchor all vertical bars.
2. Additional anchorage of foundation walls to floors usually will be required with respect to structures designed for resistance to blast or earthquake forces.

L.3 □ Design of foundation walls

Even if the load on the foundation wall can be determined with certainty, there is still the problem of calculating the strength of the wall. The load-carrying capacity of a laterally loaded masonry wall has, for known boundary conditions, been dealt with extensively in Chapter H, but the degree of restraint at the bottom end of the wall and the type of support at the upper end of the wall are often unknown. Furthermore, the effect of the cutouts for the basement windows complicates the problem. An additional complication is the anisotropic behavior of the masonry wall.

Figure L.7 □ Typical detail of a joint between a masonry foundation wall, first floor slab (or joist floor), and an external wall. According to National Concrete Masonry Association [L.6].

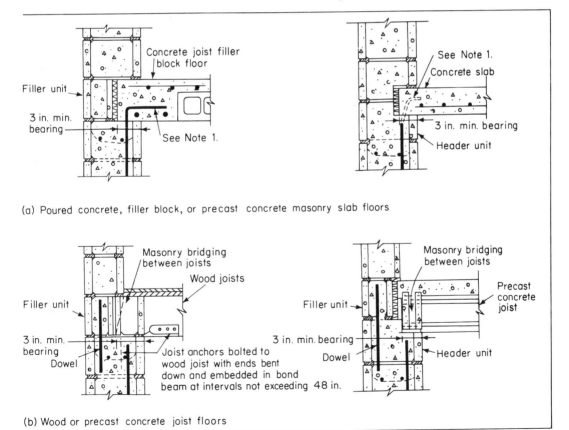

(a) Poured concrete, filler block, or precast concrete masonry slab floors

(b) Wood or precast concrete joist floors

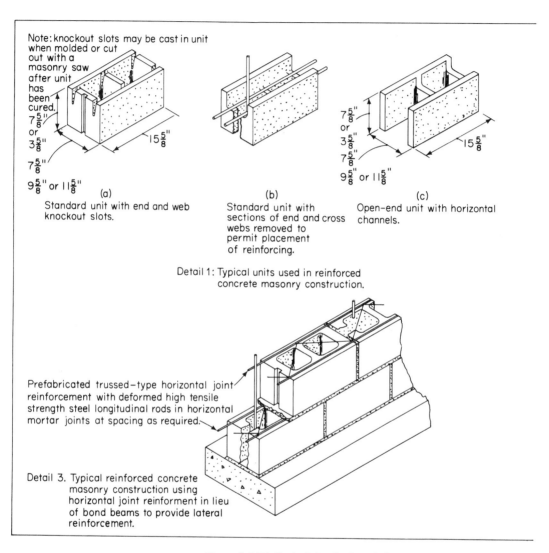

Note: knockout slots may be cast in unit when molded or cut out with a masonry saw after unit has been cured.

$7\frac{5}{8}$" or $3\frac{5}{8}$"

$7\frac{5}{8}$"

$9\frac{5}{8}$" or $11\frac{5}{8}$"

$15\frac{5}{8}$"

(a)

Standard unit with end and web knockout slots.

(b)

Standard unit with sections of end and cross webs removed to permit placement of reinforcing.

$7\frac{5}{8}$" or $3\frac{5}{8}$"

$7\frac{5}{8}$"

$9\frac{5}{8}$" or $11\frac{5}{8}$"

$15\frac{5}{8}$"

(c)

Open-end unit with horizontal channels.

Detail 1: Typical units used in reinforced concrete masonry construction.

Prefabricated trussed-type horizontal joint reinforcement with deformed high tensile strength steel longitudinal rods in horizontal mortar joints at spacing as required.

Detail 3. Typical reinforced concrete masonry construction using horizontal joint reinforment in lieu of bond beams to provide lateral reinforcement.

Figure L.8 □ Typical details for reinforcing a concrete masonry foundation wall according to National Concrete Masonry Association [L.6].

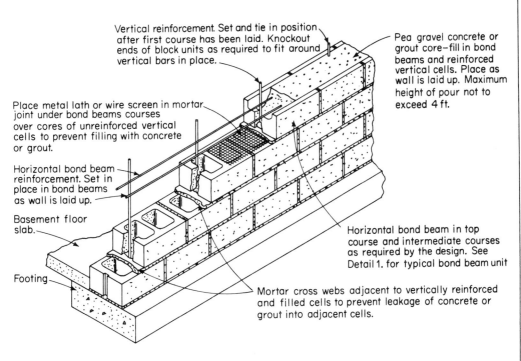

Vertical reinforcement. Set and tie in position after first course has been laid. Knockout ends of block units as required to fit around vertical bars in place.

Pea gravel concrete or grout core-fill in bond beams and reinforced vertical cells. Place as wall is laid up. Maximum height of pour not to exceed 4 ft.

Place metal lath or wire screen in mortar joint under bond beams courses over cores of unreinforced vertical cells to prevent filling with concrete or grout.

Horizontal bond beam reinforcement. Set in place in bond beams as wall is laid up.

Basement floor slab.

Footing

Horizontal bond beam in top course and intermediate courses as required by the design. See Detail 1. for typical bond beam unit

Mortar cross webs adjacent to vertically reinforced and filled cells to prevent leakage of concrete or grout into adjacent cells.

Detail 2. Typical reinforced concrete masonry construction—reinforcement and core-fill placed as wall is laid up.

One reasonable approach is to treat the wall as a slab supported along four edges and to estimate the type of restraint that is present at the different edges. From these assumptions, the moment distribution in the wall can be approximately calculated according to the elastic theory of slabs or according to the yield line theory (see textbooks on slabs [L.4], [L.5]).

A typical detailing of the foundation wall above the ground is shown in Fig. L.7 [L.6], in which the placement of vertical reinforcement is also shown. Reinforcement can be necessary in cases where the lateral loads are high and where the friction between the floor slabs and the foundation wall is too small to give a support at the upper end of the wall to transmit lateral forces from wind, etc. A typical method of reinforcing a hollow concrete block masonry wall is shown in Fig. L.8.

In extreme cases the wall must be reinforced and grouted to form a very strong reinforced composite resembling a reinforced concrete wall. The reader should also consult Chapters K and M for examples of detailing.

References for Chapter L

L.1 ☐ Terzaghi, Karl, and Peck, Ralph B.: *Soil Mechanics in Engineering Practice*. John Wiley & Sons, Inc., New York, 1967.

L.2 ☐ Huntington, Whitney C.: "Earth Pressures and Retaining Walls." John Wiley & Sons, Inc., New York, 1957.

L.3 ☐ Broms, B., and Rehnman, S. E.: "Jordtryck mot grundmurar av Lecablock." Väg-och vattenbyggaren 1.2, January–February 1968, Stockholm.

L.4 ☐ Timoshenko, S., and Woinowsky-Krieger, S.: *Theory of Plates and Shells*. McGraw-Hill Book Co., New York, 1959.

L.5 ☐ Johansen, K. W.: "The Ultimate Strength of Reinforced Concrete Slabs." Final Report, Third Congress, International Association for Bridge and Structural Engineering, Liege, September, 1948, p. 565.

L.6 ☐ National Concrete Masonry Association: "Concrete Masonry Foundation Walls." NCMA, CM 131, Arlington, Virginia, 1961.

L.7 ☐ Portland Cement Association: "Concrete Masonry Handbook for Architects, Engineers, and Builders." PCA, Skokie, Illinois, 1951.

$\underset{\text{M}}{\text{M}}$ $\square$ Codes, practical applications, and case studies

---

M.1 $\square$ Introductory remarks

Most of the values given for different properties of masonry in the foregoing chapters have been results from laboratory tests. As has been described in Chapter C, many different factors affect the strength of masonry, and, furthermore, the strength values have considerable scatter. To obtain a permissible stress from the obtained data, a safety factor must be applied. A rational approach to the choice of the magnitude of the factor would be to make a proper probabilistic calculation and from this apply different safety factors for different types of masonry and for different types of loading conditions. However, the type of information needed is normally not available. The safety factor has usually been chosen by judgement of a committee that has been in charge of a code. Although the difference in magnitude of common safety factors is considerable, the minimum seems to be about 3 and the maximum around 10 in current codes. A discussion of permissible stresses, etc., will be given in the following sections.

In addition to the requirements outlined in a code, several other important factors for designing and detailing masonry buildings should be considered. Such factors, as well as examples of structural joints, hints for location of reinforcement, and case studies, will be discussed later in this chapter.

M.2 $\square$ Codes of practice

M.2a: Introductory remarks

In the following, some of the newer codes of practice ([M.1], [M.2], [M.3], [M.4], [M.5], [M.6], [M.7], [M.8]) will be compared and discussed briefly. Comparisons of codes have also been published elsewhere ([M.9], [M.10]). Specifications for materials also have a bearing upon the use of masonry ([M.11], [M.12], [M13]). Many codes are revised frequently and if the code is used for design purposes, the latest edition must be consulted. Manufacturers' publications often furnish information about design in addition to the codes ([M.14], [M.15], [M.16]).

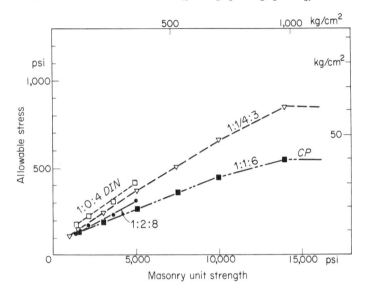

Figure M.1a ☐ Allowable stress as a function of masonry unit strength according to 1962 German (DIN) and 1964 British Codes (CP) [M.1], [M.2].

Figure M.1b ☐ Allowable stress as a function of masonry unit strength according to 1967 Swedish (BABS), 1965 Swiss (SIA), and 1967 American (A41) (SCPI) Codes [M.3], [M.4], [M.6], [M.5].

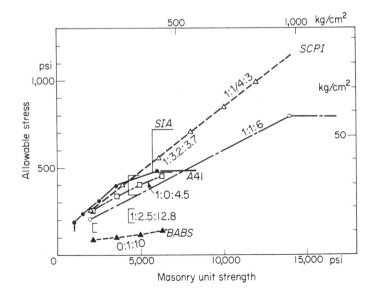

M.2b: Permissible stress for concentrically loaded walls
and columns

Two approaches to determining the strength of concentrically
loaded masonry are used in the recent codes. The prevailing
method to date is to assume the value of the masonry strength
from a given standard strength of the masonry unit and a given
type or quality of mortar. The compressive strength so obtained
is divided by a safety factor or multiplied by a coefficient which
is the inverse value of the safty factor. By this procedure a
permissible stress for a concentrically loaded wall is obtained.
The procedure contains, of course, many steps of judgments, etc.
In Figs. M.1a and M.1b the permissible stress for some combinations of unit strength and mortar mixture is shown. The
comparison contains some data from the Swedish codes, the SCPI
Code, the SIA Code, the CP 111 Code, and the DIN Code. From
each of these codes a high strength mortar and a high quality
masonry have been chosen, as well as a medium or low strength
mortar and low quality workmanship masonry. Most of the
mentioned values are valid for clay brick masonry. (Compare test
data shown in Figs. C.1a and C.1b.) A curve marked A41 has
also been plotted in Fig. M.1b. This curve applies to concrete
masonry as given in [M.6].

From the figures, it can be seen that the allowable stresses
for unit strengths over 5000 psi are between 5 and 10% of the
unit strength. For lower unit strengths the percentage is higher,
5 to 25%, depending strongly on the mortar quality.

A second, newer approach to obtaining a strength value for
concentrically loaded masonry is to make tests on masonry prisms
before the construction [M.17]. The prisms must be built of
similar materials under the same conditions and with the same
curing arrangements as for the actual structure, as far as possible.
The temperature, the moisture content, the consistency of the
mortar, and the workmanship should, as far as possible, be the
same as will be used in the actual structure. Five or more prism

Table M.1 ☐ *Safety factors and*
permissible stresses.

| | Direct stress | | Flexure (psi) | | Shear (psi) |
	Walls	Columns	⊥ bed joint	// bed joint	
SCPI	$0.25f'_m$	$0.20f'_m$	28–36	56–72	40–50
A 41	$0.20f'_m$		16–39	32–78	50

specimens are usually required to establish a mean value of the average strength. The obtained mean strength is then taken as the basic value for the masonry strength, provided that the scatter of the test results is within certain limits. In some instances the masonry strength actually obtained has also been checked on masonry prisms laid simultaneously with the walls on the building site.

The basic strength value obtained from tests (or judged from block and mortar strength) is then multiplied by a factor 0.20 or 0.25 (safety factor 5 or 4) to give the permissible stress. Some examples of the factor are shown in Table M.1.

M.2c : Reductions for high slenderness and eccentricity of axial load. Permissible stresses in bending and shear

Most codes provide a reduction coefficient for the reduction of the permissible stresses for eccentrically loaded masonry and for slender masonry when buckling is imminent. Most of the codes express the reduction as a factor less than or equal to unity to be multiplied by the permissible stress as a function of the height over thickness ratio h/d, as shown in Fig. M.2. Furthermore, a maximum slenderness ratio h/d is usually prescribed. The scatter at the higher h/d ratios is considerable, probably due to the fact that the modulus of elasticity has a great influence on the buckling strength of columns, as shown in Chapter B. Since the modulus of elasticity is not introduced as a variable in the

*Figure M.2 □ Reduction of allowable stress as a function of slenderness and load eccentricity: 1967 American (A41)(SCPI), 1968 Swiss (SIA), and 1964 British (BCP) Codes [M.6], [M.5], [M.4], [M.2].
——— is concentric load.
- - - - - is eccentric load e = d/6.*

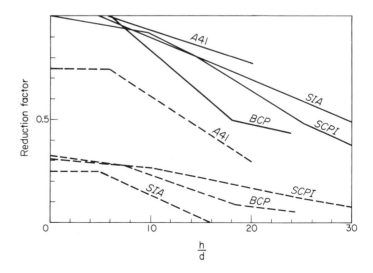

Figure M.3 □ *Reduction β of allowable stresses due to slenderness (h/d), ultimate strain (ε₁), and load eccentricity (the position of the wall in the building n = 1, 2, etc., determines the eccentricity). Example: h/d = 17, ε = 8.5, n = 2; result:* (a) *Top diagram (valid for slabs supported along all four sides):* β = 0.23. (b) *Bottom diagram (valid for simple supported slab):* β = 0.17. *(1967 Swedish codes [M.3]).*

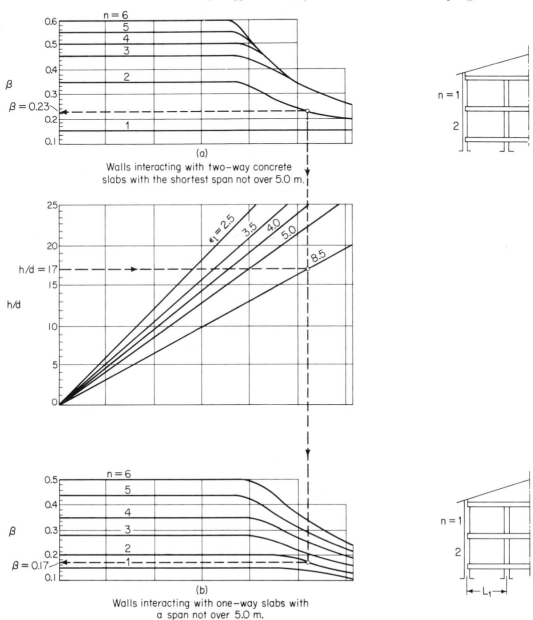

(a)
Walls interacting with two—way concrete slabs with the shortest span not over 5.0 m.

(b)
Walls interacting with one—way slabs with a span not over 5.0 m.

graph of the reduction coefficient, and since different types of masonry probably have been used in the basic test for the different codes, it is easily conceived that different values are obtained. One exception in this approach is the 1967 Swedish code, BABS, which provides graphs (Fig. M.3) for the calculation of the effect of the slenderness h/d of the wall as well as the effect of the eccentricity of the axial load. The graphs also contain a set of lines which depend upon the modulus of elasticity of the actual masonry. (Compare the buckling curves in Chapter E.)

The real reduction of the allowable stress for eccentric loading varies also, depending on the assumption upon which the calculation of the eccentricity rests. For example, a material having no tensile strength could be assumed for loading outside the kern. In this case, the stress block is triangular and a part of the cross section is cracked. In some cases, the masonry can take tensile stresses and, in such a case, a higher allowable mean stress is obtained. In some other instances, the interaction between walls and slabs is taken into account so that a statically indeterminate system is the basis for the calculation of the eccentricity of the axial load.

Most codes provide very little guidance for the calculation of the magnitude of the eccentricities of the force in the wall. However, the 1967 Swedish code (from which an example of graphs provided is shown in Fig. M.3) considers the effect of the load eccentricity. A separate graph is printed for each type of structure. The structures are classified according to the load eccentricity they produce in the walls; thereby, the reduction coefficient can be read directly from the diagrams. In cases where the structures cannot be sorted into any of the given groups, the eccentricity has to be calculated from normal procedures of structural mechanics, except that the lack of tensile strength of the masonry has to be taken into account.

The codes normally provide only one or two sets of absolute magnitudes for the permissible tensile stresses in flexure and for the permissible stresses in shear. In Table M.1, examples of permissible stresses acting perpendicular as well as parallel to the bed joints are shown. The lower values apply to a low strength mortar and the high ones to a high strength mortar type of masonry. Examples of permissible shear stresses are also shown in Table M.1. The permissible stresses for flexure and shear reflect the fact that the bond, tensile, and shear strength of the masonry depends comparatively little upon the unit strength and the mortar strength. These effects have been discussed in Chapter D. Reinforced masonry is designed with some additional considerations.

M.2d : Reinforced masonry

Most codes regulate the permissible stresses in the tensile reinforcement of masonry beams in the same way as for reinforced concrete. The bond stresses and the shear stresses are related to the actual class of masonry as determined by the types of mortar and masonry units. A minimum strength and quality of mortar and units is prescribed for reinforced masonry, to ensure corrosion protection for the steel and to guarantee interaction between the steel and the masonry.

Considerations not directly related to the strength sometimes determine the dimensions of the masonry, and although they are not directly within the scope of this book, they will be briefly mentioned.

M.2e : Fire resistance

For fire resistance, the U.S. National Building Code requires minimum face shell thickness of about $1\frac{1}{4}$ in. and web thickness of 1 in. for C-2 retardants, and more for higher fire resistance (see also ASTM E119 for type of tests).

M.2f : Thermal insulation

In some locations the thermal insulation of the wall is important. Based on tests by Rowley, Algren, and Carlson [M.18] and by Rowley, Algren, and Lander [M.19], PCA [M.14] has compiled Table M.2, showing the heat transmission U (Btu/h/ft^2/°F) for different types of walls, particularly concrete masonry walls. Table M.3 shows values compiled by Structural Clay Products Institute [M.15], particularly for clay brick masonry walls. Typical maximum values of U specified by the Federal Housing Administration in a number of cities in the U.S. in 1968 are shown in Table M.4 [M.14].

M.2g : Sound insulation

The sound insulation obtained by some typical sections of masonry walls are shown in Tables M.5 [M.14] and M.6 [M.15].

Basic wall construction*		Interior finish			
		Plain wall no plaster	$\frac{1}{2}$-in. plaster on:		
			Wall direct	$\frac{3}{4}$-in. furring with:	
				$\frac{3}{4}$-in. plasterboard	$\frac{1}{2}$-in. rigid insulation
Concrete masonry (cores not filled)	8-in. sand and gravel or limestone	0.53	0.49	0.31	0.22
	8-in. cinder	0.37	0.35	0.25	0.19
	8-in. expanded slag, clay or shale	0.33	0.32	0.23	0.18
	12-in. sand and gravel or limestone	0.49	0.45	0.30	0.22
	12-in. cinder	0.35	0.33	0.24	0.18
	12-in. expanded slag, clay or shale	0.32	0.31	0.23	0.18
Concrete masonry (cores filled with insulation)**	8-in. sand and gravel or limestone	0.39	0.37	0.26	0.19
	8-in. cinder	0.20	0.19	0.16	0.13
	8-in. expanded slag, clay or shale	0.17	0.17	0.14	0.12
	12-in. sand and gravel or limestone	0.34	0.32	0.24	0.18
	12-in. cinder	0.20	0.19	0.15	0.13
	12-in. expanded slag, clay or shale	0.15	0.14	0.12	0.11
Cavity walls (with 2-in. or larger cavity. Cavity not filled with insulation)	10-in. wall of two 4-in. sand and gravel or limestone units	0.34	0.33	0.24	0.18
	10-in. wall of two 4-in. cinder, expanded slag, clay or shale units	0.26	0.24	0.19	0.15
	10-in. wall of 4-in. face brick and 4-in. sand and gravel or limestone unit	0.38	0.36	0.25	0.19
	10-in. wall of 4-in. face brick and 4-in. cinder, expanded slag, clay or shale unit	0.33	0.31	0.23	0.18
	14-in. wall of 4-in. face brick and 8-in. sand and gravel or limestone unit	0.33	0.31	0.23	0.18
	14-in. wall of 4-in. face brick and 8-in. cinder, expanded slag, clay or shale unit	0.26	0.25	0.19	0.16
	14-in. wall of 4-in. and 8-in. sand and gravel or limestone unit	0.30	0.28	0.21	0.17
	14-in. wall of 4-in. and 8-in. cinder, expanded slag, clay or shale unit	0.22	0.21	0.17	0.14

M.2h: Weather resistance

Freezing and thawing and wind-driven rain are the main weather factors affecting masonry. The weather resistance of masonry units can be judged by some standard tests ([M.11], [M.12]), but the resistance of the masonry is greatly dependent upon complete filling of mortar joints. The resistance against driving rain especially relies on the tightness of the joints.

Table M.2 □ *Coefficients of heat transmission (U) for various walls.*
*All concrete masonry shown in this table are hollow units. All concrete masonry wall surfaces exposed to the weather have two coats of Portland cement base paint. Surfaces of all walls exposed to the weather subject to a wind velocity of 15 miles per hour. **Values based on dry insulation. The use of vapor barriers or other precautions must be considered to keep insulation dry.*

Basic wall construction*		Plain wall no plaster	Interior finish		
			Wall direct	½-in. plaster on:	
				¾-in. furring with:	
				¾-in. plaster-board	½-in. rigid insulation
4-in. face brick plus:	4-in. sand and gravel or limestone unit	0.53	0.49	0.31	0.23
	4-in. cinder, expanded slag, clay or shale unit	0.44	0.42	0.28	0.21
	4-in. common brick	0.50	0.46	0.30	0.22
	8-in. sand and gravel or limestone unit	0.44	0.41	0.28	0.21
	8-in. cinder, expanded slag, clay or shale unit	0.31	0.30	0.22	0.17
	8-in. common brick	0.36	0.34	0.24	0.19
	1-in. wood sheathing, paper, 2 × 4 studs, wood lath and plaster	—	0.27	0.27	0.20
4-in. common brick plus:	4-in. sand and gravel or limestone unit	0.45	0.42	0.28	0.21
	4-in. cinder, expanded slag, clay or shale unit	0.38	0.36	0.26	0.19
	8-in. sand and gravel or limestone unit	0.37	0.35	0.25	0.19
	8-in. cinder, expanded slag, clay or shale unit	0.28	0.27	0.20	0.16
	8-in. common brick	0.31	0.30	0.22	0.17
	1-in. wood sheathing, paper, 2 × 4 studs, wood lath and plaster	—	0.25	0.25	0.19
Wood frame	wood siding, 1-in. wood sheathing, 2 × 4 studs, wood lath and plaster	—	0.25	0.24	0.19

M.2i: Concluding remarks about codes

As a general comment, in can be said that most of the codes permit relatively low permissible stresses compared with the strength of the masonry under laboratory conditions. By increased quality control of materials and workmanship, there is room for better utilization of load-bearing masonry.

In addition to the structural strength and safety of masonry that have been dealt with in the foregoing, other technical and

Table M.3 ☐ Coefficients of heat transmission (U) and resistances (R) of various clay masonry walls. Computed for plain walls with no interior finishes. (Corrected for wind velocity of 15 miles per hour.)

Wall number	Wall type	$\left(\dfrac{U}{\frac{1}{R}}\right)$	$\left(\dfrac{R}{\frac{1}{U}}\right)$
1	6″ solid brick ("SCR brick")	0.72	1.38
2	8″ solid brick	0.55	1.82
3	12″ solid brick	0.43	2.34
4	8″ tile (2 cell, flat bed)	0.41	2.45
5	8″ tile (3 cell, flat bed)	0.36	2.80
6	8″ tile (3 cell, recessed bed)	0.31	3.28
7	12″ tile (2 unit wall, 4″ and 8″, with latter same as No. 4)	0.29	3.48
8	12″ tile (2 unit wall, 4″ and 8″, with latter same as No. 5)	0.26	3.83
9	12″ tile (2 unit wall, 4″ and 8″, with latter same as No. 6)	0.23	4.31
10	8″ brick and tile	0.43	2.33
11	12″ brick and tile (tile same as No. 4)	0.33	2.89
12	12″ brick and tile (tile same as No. 5)	0.31	3.24
13	12″ brick and tile (tile same as No. 6)	0.27	3.72
14	10″ brick cavity	0.36	2.77
15	10″ tile cavity	0.26	3.85
16	10″ brick and tile cavity	0.31	3.24
17	14″ tile cavity (inside wythe same as No. 4)	0.23	4.39
18	14″ tile cavity (inside wythe same as No. 5)	0.21	4.74
19	14″ tile cavity (inside wythe same as No. 6)	0.19	5.22
20	14″ brick and tile cavity (inside wythe No.4)	0.26	3.88
21	14″ brick and tile cavity (inside wythe same as No. 5)	0.24	4.23
22	14″ brick and tile cavity (inside wythe same as No. 6)	0.21	4.71

*Table M. 4 ☐ Typical FHA heat transmission coefficient limitations for walls for one or two living units. *From section 402-A of FHA Minimum Property Requirements for Properties of One or Two Living Units for various insuring offices, July 1948.*

FHA insuring office	Overall coefficient of heat transmission (U factor)*
Arizona (Pheonix)	0.50
Florida (Jacksonville)	0.33
Illinois (Chicago)	0.27
Montana (Helena)	0.13
New York (New York)	0.30
Oregon (Portland)	0.25
Minnesota (Minneapolis)	0.19
Texas (Dallas)	0.35

Table M.5 □ Reduction factors in sound transmission through walls of hollow concrete masonry. [a]National Bureau of Standards Report BMS17. [b]Data reported in Acoustics and Architecture by Paul E. Sabine. [c]Tests conducted at Riverbank Laboratories. [d]National Bureau of Standards Supplement to Report BMS17.

Walls of hollow concrete masonry	Weight per sq. ft. of wall area, lb.	Average reduction factor, decibels
3″ Cinder, $\frac{5}{8}$″ plaster on both sides[a]	32.2	45.1
4″ Cinder, $\frac{5}{8}$″ plaster on both sides[a]	35.8	45.6
4″ Cinder, 1″ plaster[b]	32.3	47.0
8″ Expanded slag, 1″ plaster[b]	56.0	52.6
4″ Celocrete, $\frac{1}{2}$″ plaster on both sides[c]	30.0	42.6
8″ Celocrete, unplastered[c]	28.6	43.7
8″ Celocrete, $\frac{1}{2}$″ plaster on both sides[c]	40.0	52.9
Cavity wall, two 4″ Celocrete, $\frac{1}{2}$″ plaster on one inner face[c]	45.0	57.1
3″ Haydite, unplastered[c]	——	36.0
3″ Haydite, 1″ plaster[c]	——	42.0
4″ Haydite, unplastered[c]	——	37.0
4″ Haydite, 1″ plaster[c]	——	43.0
6″ Haydite, unplastered[c]	——	44.8
6″ Haydite, 1″ plaster[c]	——	48.5
8″ Haydite, unplastered[c]	——	47.8
8″ Haydite, 1″ plaster[c]	——	50.5
12″ Haydite, unplastered[c]	——	52.0
12″ Haydite, 1″ plaster[c]	——	54.0
4″ Pumice, $\frac{1}{2}$″ plaster on both sides[d]	25.3	37.4
4″ Pumice, $\frac{1}{2}$″ plaster on one side only	20.4	34.6
4″ Waylite, $\frac{1}{2}$″ plaster on both sides[c]	31.0	50.0
8″ Waylite, $\frac{1}{2}$″ plaster on both sides[c]	47.0	53.0
3″ Waylite, 2 coats cement paint each side[c]	16.75	44.1
4″ Waylite, unpainted[c]	16.5	33.2
4″ Waylite, 2 coats cement paint each side[c]	16.5	46.7
6″ Waylite, unpainted[c]	21.0	39.7
6″ Waylite, 2 coats cement paint each side[c]	21.0	52.2
Cavity wall, two 3″ Waylite, $\frac{3}{8}$″ plaster on one unexposed face[c]	17.0	56.1

economical problems could deserve consideration in some instances and are in fact regulated to some extent in most codes. Thermal insulation and sound insulation can sometimes determine a wall thickness. Sometimes the desired masonry unit type can be too expensive and a local unit must be used with corresponding adjustments in mortar quality and/or wall thickness, etc. Fire resistance can be of primary concern in some instances. Sometimes the masonry type is chosen on an esthetic basis and sometimes by virtue of tradition. Some other factors of importance for the practical application are discussed briefly in the following sections.

Table M.6 ☐ Sound transmission loss in decibels of various clay masonry walls.

Type of wall	Weight psf	Average reduction factor decibels
12″ hollow tile (2 units in wall thickness), plastered both sides with brown coat and white finish	66	50.0
8″ hollow tile, plastered both sides with brown coat and smooth white finish	48	49.8
8″ hollow tile (Heath cubes) plastered both sides with brown coat and smooth white finish	55	51.0
6″ hollow tile, plastered both sides with brown coat and smooth white finish	39	47.1
4″ hollow tile, plastered both sides with brown coat and smooth white finish	29	44.0
3″ hollow tile plastered both sides with brown coat and smooth white finish	28	44.4
4″ hollow tile, wood furring, paper, metal lath and gypsum plaster both sides	34	57.5
$2\frac{1}{4}$″ brick, plastered both sides	31.6	48.8
$2\frac{1}{4}$″ brick, furring strips, gypsum plaster board and plaster both sides	38.2	55.2
$2\frac{1}{4}$″ brick, furring strips, $\frac{1}{2}$″ Insulite and plaster both sides	33.3	54.6
4″ brick, gypsum plaster with smooth lime finish both sides	49	53.7
8″ brick, plastered both sides with brown coat and white finish	87	57.2

M.3 ☐ Planning, detailing, and construction techniques

M.3a: Introductory remarks

The quality of the masonry is determined at the design office and on the construction site. In the following, some practical rules for planning, detailing, and construction are given.

M.3b: Planning and detailing

A number of important factors must be taken into account in the layout of plans and patterns for a masonry structure. (1) The sizes of the rooms must be adjusted to fit the sizes of the masonry units and joints. This can be accomplished with a modular system. (2) Window and door openings must be placed according to the modular mesh to avoid excessive numbers of nonstandard formats of masonry units. (3) Nonstandard formats should be fabricated and used whenever necessary. For example, quarter, half, and three-quarter units should always be used instead of split units. (4) Cross walls should be located where they can give support to highly stressed portions of walls, thereby decreasing the risk of buckling. (5) The bed joints should be oriented perpendicular to the axis of the forces to provide the highest possible strength. (6) The walls should be loaded so that the strains are as nearly equal as possible in all walls in a horizontal section of a building, in order to avoid differential vertical movements. (7) The walls should be placed concentrically on top of each other over the whole height of the building in order not to induce moments at the joints. (8) Whenever possible, the detailing should aim at a concentric transfer of load from slabs to walls. (9) No chasing of masonry must be permitted. Plastic-coated cables can be used instead of the normal electrical conduits. The cables can be placed on the walls and plastered over. All plumbing must be put into vertical ducts provided in the building plans. (10) It is advisable to reinforce the masonry around large openings in highly stressed masonry to prevent cracking. (11) Cracking in low stressed walls due to shrinkage and deflection of slabs, especially at the top of the building, as well as uplift of the corner of the slabs must be prevented with proper measures or covered over to provide a good appearance. (12) A maximum permissible deviation from given locations of walls must be prescribed, and a maximum deviation from straightness must be required (3 to 4 mm seems to be a reasonable value). (13) A maximum joint thickness must be prescribed. (14) In all external faces of the masonry, the joints should be tuckpointed, in order to prevent penetration by rain or moisture. (15) A minimum temperature at which work can proceed must also be prescribed. (16) National and local codes must be complied with. (17) Manufacturers publications should be consulted. (18) Availability of units should be checked locally.

M.3c : Construction techniques

The quality of the final product depends heavily upon the techniques used in the actual construction. An adoption of the following rules will aid in obtaining a good quality structural masonry: (1) All joints must be completely filled. (2) Clay bricks with high suctions should be wetted (but not overwetted) before laying (to reduce the detrimental effects of high suction on the strength properties). (3) Concrete masonry blocks should not be laid wet, because the drying shrinkage probably will be

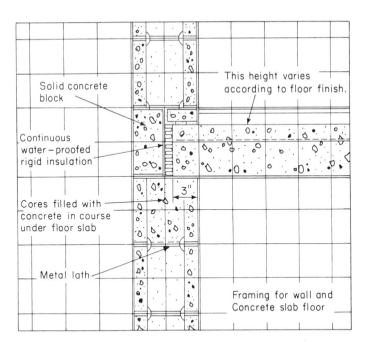

This height varies according to floor finish.

Solid concrete block

Continuous water-proofed rigid insulation

Cores filled with concrete in course under floor slab

3"

Metal lath

Framing for wall and Concrete slab floor

Figure M.4 ☐ Typical connection between concrete masonry block walls and concrete slabs. National Concrete Masonry Association [M.14].

excessive. (4) A high flow mortar with high water retentivity should be used. (5) Shields against weather should be provided when weather conditions would otherwise prevent the masons from complying with the rules.

M.4 ☐ Joints with slabs

Figure M.4 shows a typical detailing of the joint between a wall and a concrete slab floor [M.14]. A joint between a brick masonry wall and wood joist floor is shown in Fig. M.5; in Fig. M.6 the joint between a cast-in-place concrete foundation wall and slab is shown. The wall placed on top of the foundation wall is a brick masonry cavity wall insulated with mineral wool. The detail shown in Fig. M.6 is designed for good thermal insulation.

M.5 ☐ Joints with roof structure

A light roof structure must be anchored to the wall structure. A method of doing this is shown in Fig. M.7 [M.14]. This type of solution can be employed in concrete block masonry struc-

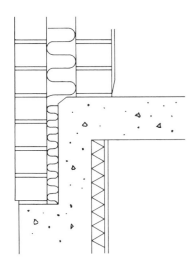

Figure M.6 □ Typical connection between concrete foundation wall, concrete slab, and brick masonry cavity wall, thermal insulation in basement and in the cavity (Nevander [M.26]).

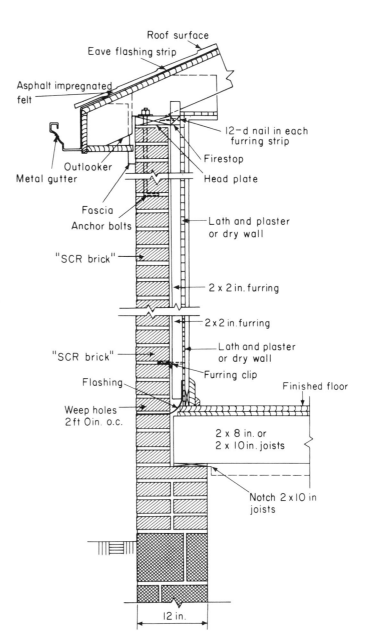

Figure M.5 □ Typical connection between masonry foundation wall, wood joist floor, and external brick masonry wall with internal plaster or dry wall. Typical connection between an external brick masonry wall and a wooden roof structure. (Structural Clay Products Institute [M.5]).

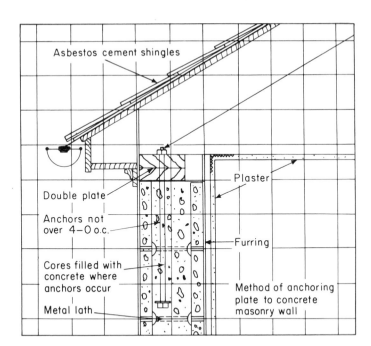

Asbestos cement shingles

Double plate

Anchors not over 4-0 o.c.

Cores filled with concrete where anchors occur

Metal lath

Plaster

Furring

Method of anchoring plate to concrete masonry wall

Figure M.7 ☐ Typical connection between external concrete hollow block masonry wall and wooden roof structure (National Concrete Masonry Association [M.14]).

tures. A similar type of anchorage must also be used in brick masonry, as shown in Fig. M.5. The anchor bolt can be omitted only in cases where the roof structure itself is heavy enough to take the uplift from wind loads or can be anchored in building parts other than the walls. Some kind of fasteners, nails, etc., should always be provided in order to keep the plates in proper position. When the external finish of the wall is likely to show eventual cracking, a sliding support of plastics or rubber could minimize the risk of architectural damage due to roof expansions.

M.6 ☐ Reinforcing

The placement of reinforcing bars in masonry sometimes causes difficulties, since the bars can be placed only in joints or in holes in the masonry units. With careful planning of the masonry pattern, however, the difficulties can be overcome. For example, hollow blocks can be placed over reinforcement by lifting them over the vertical bar ends; the horizontal bars can be placed in the horizontal joints. Open-end units can be placed directly over vertical bars, as shown in Fig. L.8. The vertical bars can also be placed afterward and grout can be poured into the holes in the

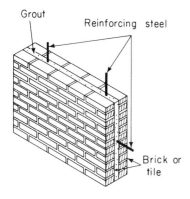

Grout

Reinforcing steel

Brick or
tile

Figure M.8 ☐ *Typical arrangement of reinforcing steel in brick masonry wall with grouted cavity (Structural Clay Products Institute [M.15]).*

blocks. The reinforcing steel can be placed in a cavity which later is filled with grout, as exemplified in Fig. M.8. Heavy horizontal reinforcement, as for example at the lower edge of a beam or at the top of a masonry wall, can be placed in units with horizontal channels. Portions of the units can also be knocked out to provide space for the bars. This procedure should, however, be used only as a last resort. Mortar joints must have sufficient thickness to accommodate the reinforcement bars plus some mortar to prevent the bars from corroding and to give a good bond between the steel bars and the masonry.

M.7 ☐ Expansion joints

Expansion joints must be made very soft and through the entire thickness of the wall, in order to accommodate the movements accumulated over the unbroken length of the adjacent wall portions. Examples of expansion joints are shown in Fig. M.9 [M.15].

Cavity walls can show considerable differential movements between the outer wythe and the inner, and the ties must be flexible enough to permit this movement without overstress, yielding, or fatigue. In the 16-story apartment buildings near Biel, Switzerland, described by Haller [M.20] (see M.9b below), in which the exterior walls are load-bearing cavity walls, the external wythe is anchored to the rest of the building only at the floor levels, with stainless steel ties permitting the external wythe to move freely during expansion due to heating and cooling. The separation of the outside wythe and the inside wythe is complete all the way around the building, even at window jambs, so that the wood casing may slip up and down.

M.8 ☐ Tolerances

According to experiences reported by Haller [M.20], the output of the masons may be considerably increased by improving the tolerances of the bricks. Due to the increase in efficiency, the overall costs are decreased as the brick quality is increased.

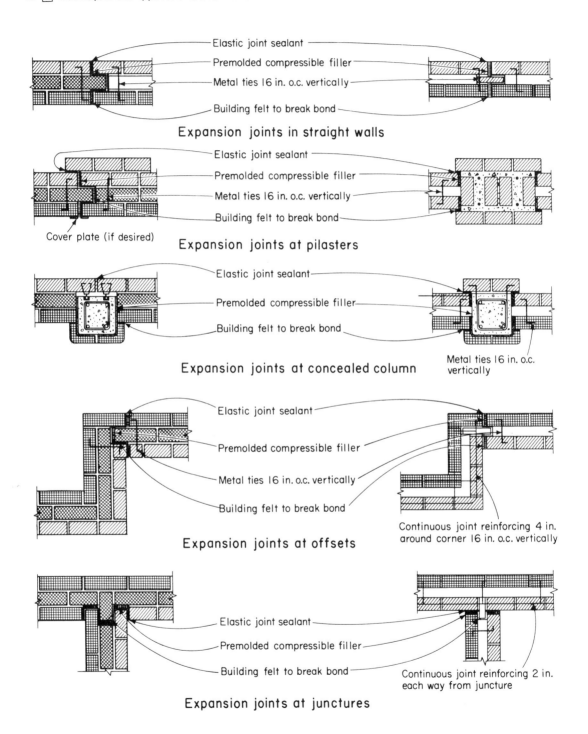

Elastic joint sealant
Premolded compressible filler
Metal ties 16 in. o.c. vertically
Building felt to break bond

Expansion joints in straight walls

Elastic joint sealant
Premolded compressible filler
Metal ties 16 in. o.c. vertically
Building felt to break bond

Cover plate (if desired)

Expansion joints at pilasters

Elastic joint sealant
Premolded compressible filler
Building felt to break bond

Expansion joints at concealed column

Metal ties 16 in. o.c. vertically

Elastic joint sealant
Premolded compressible filler
Metal ties 16 in. o.c. vertically
Building felt to break bond

Expansion joints at offsets

Continuous joint reinforcing 4 in. around corner 16 in. o.c. vertically

Elastic joint sealant
Premolded compressible filler
Building felt to break bond

Continuous joint reinforcing 2 in. each way from juncture

Expansion joints at junctures

M.9 ☐ Case studies

M.9a: Introductory remarks

High buildings of load-bearing masonry are not new. In 1891, a 16-story building with load-bearing walls (the Monadnock Building [M.21]) was built in Chicago. Older buildings had very low permissible stresses in the masonry. For example, the Monadnock Building has about 6-ft-thick walls at the base.

The new concept in engineered masonry is the recognition of masonry as a structural material just like any other building material. The strength of the masonry can be utilized to carry loads in consort with other structural parts of the building, such as walls and floors of concrete, steel, or wood. At the same time,

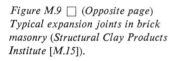

Figure M.9 ☐ (Opposite page) Typical expansion joints in brick masonry (Structural Clay Products Institute [M.15]).

Figure M.10 ☐ (Right) Plan of modern apartment building in Switzerland, 1951. The building has thirteen stories. [M.22]. Architects: Gfeller and Mähli. Structural engineer: Geering.

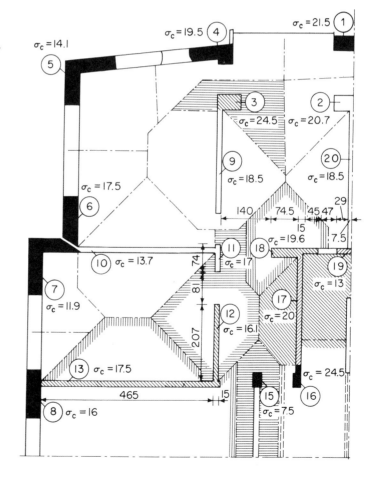

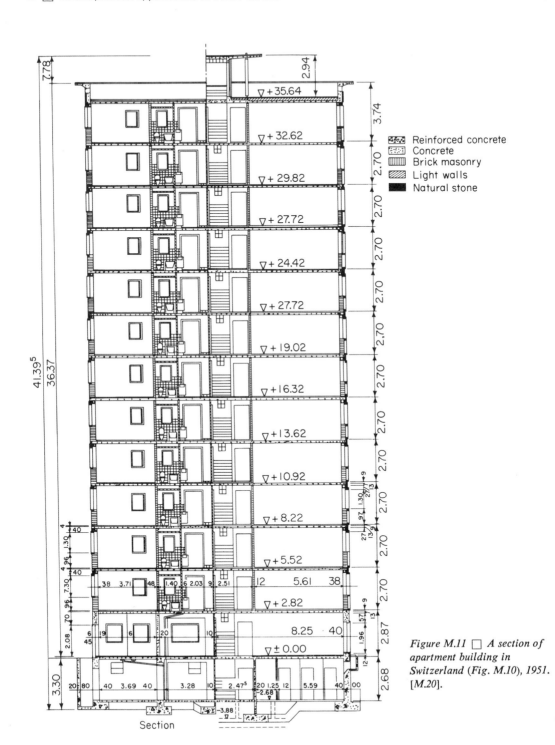

Reinforced concrete
Concrete
Brick masonry
Light walls
Natural stone

Section

Figure M.11 □ A section of apartment building in Switzerland (Fig. M.10), 1951. [M.20].

such uses require more precise knowledge of the strength of the masonry under different conditions, as well as better control on the building site and in the factory. The fabrication of masonry units has been greatly improved in the last few decades by mechanization, which provides a final product of predictable quality. The calculation methods, building codes, and transportation techniques have improved substantially, permitting a better and more economical utilization of masonry as a load-bearing structural element. A few recent achievements in this field are briefly described below.

M.9b: Swiss 13-story apartment buildings: Basel, 1951

The first modern high rise brick buildings are believed to be the three 13-story apartment buildings in Switzerland which use the masonry as a load-bearing structure element. The plans of the building are shown in Fig. M.10, taken from Rilem [M.22]. The maximum interior wall thickness is 6 in. The external walls are at most 15 in. thick. None of the load-bearing walls were reinforced. The maximum brick strength was 300 kg/cm² (4300 psi). The building process and the accompanying experiments are described in detail by Haller [M.20]. A section of the building is shown in Fig. M.11 and a photograph of the exterior in Fig. M.12.

Figure M.12 □ A photograph of apartment building in Switzerland (Fig. M.10), 1951. [M.20].

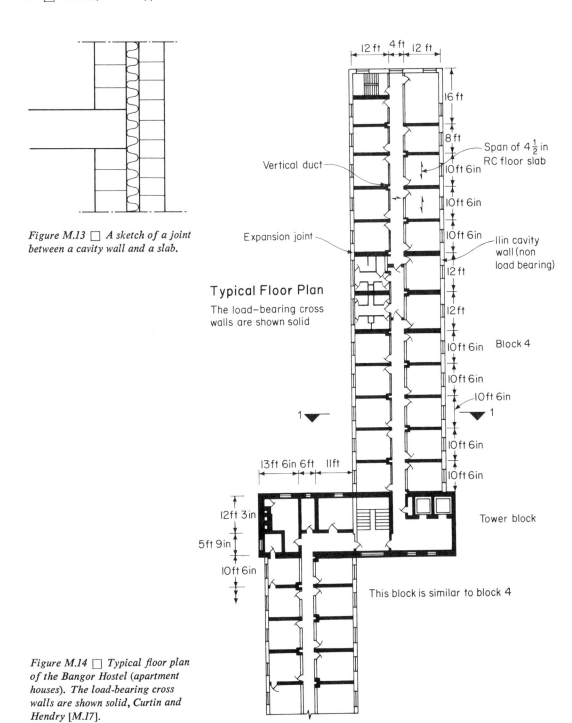

Figure M.13 ☐ A sketch of a joint between a cavity wall and a slab.

Typical Floor Plan

The load-bearing cross walls are shown solid

Figure M.14 ☐ Typical floor plan of the Bangor Hostel (apartment houses). The load-bearing cross walls are shown solid, Curtin and Hendry [M.17].

Vertical duct

Expansion joint

Span of $4\frac{1}{2}$ in RC floor slab

11in cavity wall (non load bearing)

Block 4

Tower block

This block is similar to block 4

12 ft · 4 ft · 12 ft

16 ft
8 ft
10 ft 6in
10 ft 6in
10 ft 6in
12 ft
12 ft
10 ft 6in
10 ft 6in
10 ft 6in
10 ft 6in
10 ft 6in

13ft 6in · 6ft · 11ft

12ft 3in
5ft 9in
10ft 6in

M.9c: Apartment buildings in Biel, Switzerland

The Biel apartment houses in Switzerland [M.23] had a maximum thickness of the load-carrying walls of 6 in. and the total height of the building was 16 stories. The total thickness of the external wall was about 15 in.

The external wythe was tied to the load-carrying internal wythe only at the level of each slab, thus permitting the external wythe to move easily due to thermal expansion, shrinkage, and creep. Furthermore, the cavity was used for thermal insulation, which consisted of about $1\frac{1}{2}$-in. thick mineral-wool mat. A sketch of the joint between the cavity wall and the slab can be seen in Fig. M.13.

Figure M.15 □ Section through the Bangor Hostel masonry houses, Curtin and Hendry [M.17]. The different brick strengths and the different mortar mixes are shown on the section.

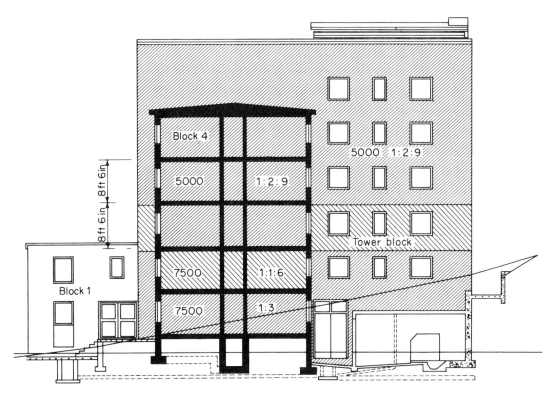

Section 1 — 1

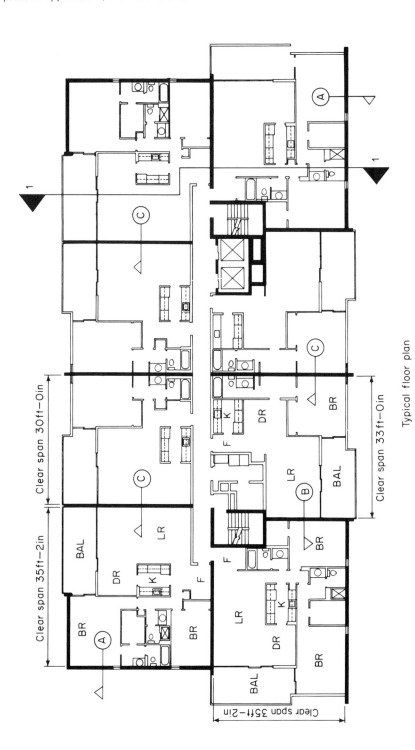

Figure M.16 ☐ Typical floor plan of Park Mayfair East in Denver, Colorado, Structural Clay Products Institute [M.24]. Architects: Anderson & Looms, Structural engineers: Sallada & Hanson.

M.9d: Six-story residential halls: University College, Bangor, North Wales

The design and construction of the Bangor Hostels is described in a paper by Curtin and Hendry [M.17]. A plan and a section of the building are given in Figs. M.14 and M.15. The maximum height of the building is six stories and the load-carrying masonry walls have a minimum thickness of $4\frac{1}{2}$ in. Bricks of two different strengths were used. The high strength bricks were used in the lower stories of the building. Furthermore, three different mixes were used: 1:0:3 for the foundation up to the first story; 1:1:6 for the second story, and 1:2:9 for the third, fourth, and fifth stories. The quality of the masonry was checked at the building site, where masonry prisms were made from the material used at the time of the sampling. The results from the crushing of these prisms were compared with similar prisms prepared under controlled conditions. By this type of control, a high and even standard of brick laying was consistently maintained.

M.9e: Earthquake-resistant 17-story masonry building

As an example of an earthquake-resistant reinforced masonry building, we cite Park Mayfair East, in Denver, Colorado [M.24]. A typical floor plan is shown as Fig. M.16 and a section of the building as Fig. M.17. Some typical wall slab connections are shown in Fig. M.18. Due to the earthquake forces (zone 1), the emphasis is laid on continuity at the joints. The top reinforcements in the double-T slabs are bent into the grouted core in the external walls or continue through the wall at internal joints. The allowable stress on the masonry wall was 50 kg/cm² (675 psi) in axial compression, and 63 kg/cm² (900 psi) in flexural compression. The overall thickness of the load-bearing masonry walls is 11 in. The grout, composed of Portland cement, sand, and pea gravel in the proportions 1:3:2, had a slump of 9 in. and was pumped into the cavities as shown in Fig. M.19. The cavity was filled in two lifts (one 4 ft and the other $4\frac{1}{2}$ ft). The grout was vibrated with a $\frac{3}{4}$-in. diameter vibrator, and after 30 to 45 min it was reconsolidated by additional vibration.

M.9f: Concrete block masonry building, New Zealand

A photograph of apartment houses in Millbrook, New Zealand, designed by Holmes [M.25] is shown as Fig. M.20. A typical section of the load-bearing external wall is shown in Fig. M.21.

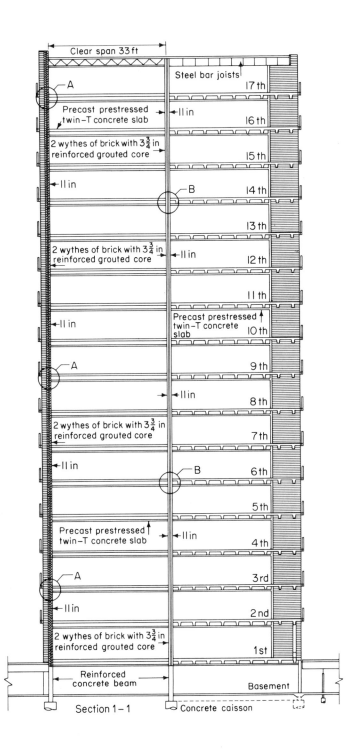

Clear span 33 ft

Steel bar joists

A

Precast prestressed twin–T concrete slab ←11 in

2 wythes of brick with 3¾ in reinforced grouted core

←11 in

2 wythes of brick with 3¾ in reinforced grouted core ←11 in

←11 in

B

Precast prestressed twin–T concrete slab

A

←11 in

2 wythes of brick with 3¾ in reinforced grouted core

←11 in

B

Precast prestressed twin–T concrete slab ←11 in

A

←11 in

2 wythes of brick with 3¾ in reinforced grouted core

17 th
16 th
15 th
14 th
13 th
12 th
11 th
10 th
9 th
8 th
7 th
6 th
5 th
4 th
3 rd
2 nd
1 st

Basement

Reinforced concrete beam

Section 1–1   Concrete caisson

Figure M.17 ☐ Cross section of load-bearing masonry houses in Park Mayfair East, Denver, Colorado, Structural Clay Products Institute [M.24].

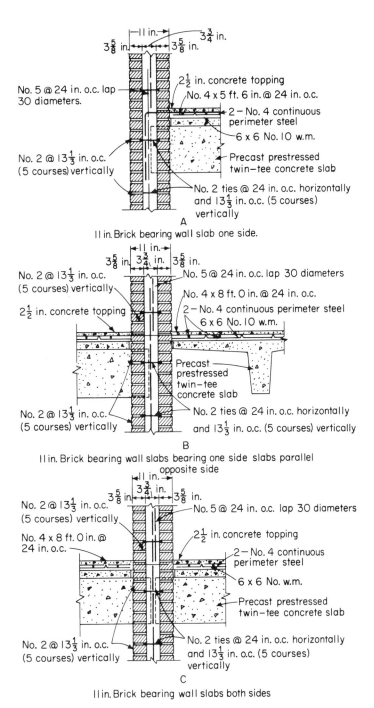

A

11 in. Brick bearing wall slab one side.

B

11 in. Brick bearing wall slabs bearing one side slabs parallel opposite side

C

11 in. Brick bearing wall slabs both sides

Figure M.18 □ *Typical details of joints between prestressed reinforced double-T concrete slabs and the reinforced grouted cavity wall, Structural Clay Products Institute* [*M.24*].

Figure M.19 □ (Above) Photograph of the grouting of the reinforced cavities of the load-bearing reinforced brick masonry walls in Park Mayfair East, Denver, Colorado, Structural Clay Products Institute [M.24].

Figure M.20 □ (Below) Photograph of apartment houses in Millbrook, New Zealand. Nine-story load-bearing reinforced concrete block masonry walls, Holmes [M.25]. Architects: Donnithorne, Structural engineers: Holmes & Wood.

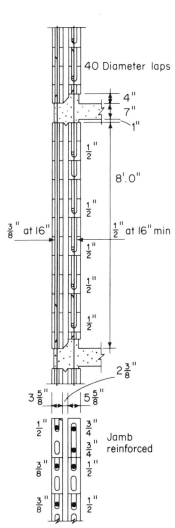

Figure M.21 □ (Above) Cross section of reinforced cavity wall for the Millbrook apartment houses in New Zealand. No grouting, no ties in the cavity. The internal wythe is thicker to accommodate the thicker reinforcing bars, Holmes [M.25].

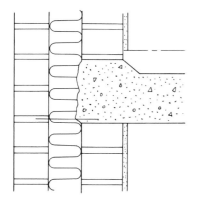

Figure M.22 □ (Above) Section of thermally insulated cavity brick masonry wall and cast-in-place reinforced concrete slabs [M.26].

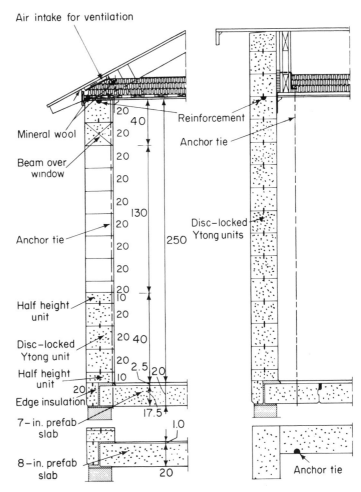

Air intake for ventilation

Mineral wool

Beam over window

Anchor tie

Reinforcement

Anchor tie

Disc-locked Ytong units

Half height unit

Disc-locked Ytong unit

Half height unit

Edge insulation

7-in. prefab slab

8-in. prefab slab

Anchor tie

Figure M.23 □ (Below) Detail of window opening. The window sash is fixed to the internal wythe and the joint between the external wythe and the window sash is filled with elastic compound to permit movement of the external wythe due to thermal expansion, settlement, etc., Nevander [M.26].

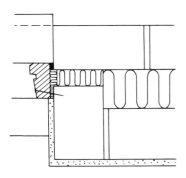

Figure M.24 □ (Above) Cross section through an external wall of disc-locked light weight cellular concrete blocks (Ytong) in a one-story house. The light wooden external roof is anchored at all four corners of a building. In larger houses, anchorage must be provided also between the corners. A typical anchor is shown in the figure. The steel bar is hidden in a chasing in the wall, according to the manufacturer [M.27].

Figure M.25 ☐ Close-up of the placement of plastic disc in the grooves in Ytong blocks [M.28].

The building is nine stories high. The load-bearing walls are of concrete block masonry with vertical and horizontal reinforcements as shown in Fig. M.21. The total thickness of the wall is $11\frac{5}{8}$ in. The two wythes of the wall are held together only at the floor levels. No ties were used. The internal wythe is made thicker in order to better accommodate the vertical reinforcement. (The design earthquake forces on New Zealand are 5 to 10% of the gravity load; 5% for buildings with natural frequencies of over 1.8 cycles per second; 10% under 0.45 cycles per second and a linear interpolation in between.) The maximum permissible stress in a reinforced masonry wall is 400 psi for flexural compression.

M.9g : Cavity wall with high thermal insulation

A type of cavity wall with high thermal insulation of mineral wool has been used extensively in Sweden in recent years. The heat flow through the wall is 0.31 kcal/m²h°C. A typical cross section of the wall at the level of floor slab is shown in Fig. M.22, taken from Nevander [M.26]. A typical detail of the connection between the window panes and the wall is shown in Fig. M.23. The window is fastened to the internal wythe only, thereby permitting the external wythe to move due to thermal expansion. The joint between the window pane and the external wythe is filled with an elastic compound. This type of wall is being modified to fit the new modular system so that the bricks have the size $9 \times 9 \times 29$ cm, fitting into a modular mesh of 10 cm.

M.9h : Masonry with no tensile strength

As an example of cases where the theory for walls without tensile strength is closest to reality, Fig. M.24 is shown [M.27]. This figure shows a typical section of an external load-bearing wall of light weight cellular concrete (Ytong). The blocks are held together by small plastic discs placed in precut grooves as shown in Figs. M.25 and M.26 [M.28]. In addition, the axial compressive force holds the blocks together, as outlined in Chapter E. One of the main advantages of this type of block is that it can be used in temperatures below freezing. Furthermore, no skill in handling mortar is required. This type of masonry block (and tongue and groove units, Siporex) are extensively used in Sweden for one- to three-story buildings. (Three stories is the maximum permissible building height without lift.) The only technical

Figure M.26 □ *Photograph of block laying in winter time for small homes in Sweden. The units are placed without mortar* [*M.28*].

limitation on the height of the building, however, is the permissible compressive stress of the masonry. The permissible stresses are 4 to 13.5 kg/cm² (55 to 190 psi), depending upon the quality of the blocks. The lighter blocks have lower permissible stresses but higher thermal insulation.

M.9i: Load-bearing masonry elements

Figure M.27 shows an example of load-bearing brick masonry cavity wall elements. The element, which is described by Elgenstierna [M.29], consists of two wythes of brick masonry plus 10-cm (4-in.) thick mineral wool mat for thermal insulation. The external wythe is 6 cm thick ($2\frac{1}{3}$ in.) and the internal wythe is 12 cm ($4\frac{3}{4}$ in.), which makes the total thickness 28 cm (11 in.). The maximum height of apartment houses which can be built of these load-bearing elements is claimed to be 8 to 10 stories.

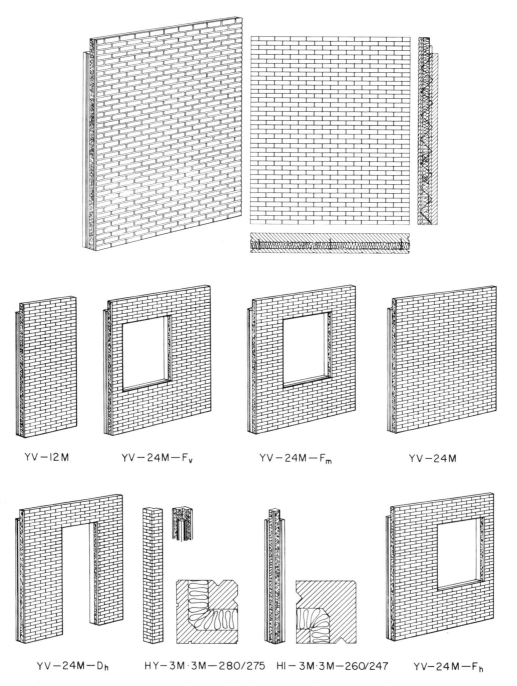

YV-12M YV-24M-F$_v$ YV-24M-F$_m$ YV-24M

YV-24M-D$_h$ HY-3M·3M-280/275 HI-3M·3M-260/247 YV-24M-F$_h$

Figure M.27 □ Typical brick masonry prefabricated element for external walls used in Sweden. The two wythes of brick masonry are connected with reinforcement of stainless steel and the two wythes are separated by thermal insulation of mineral wool, according to Elgenstierna [M.29].

Figure M.28 □ *Photograph of office building for Sydkraft [M.30].*

M.9j : "Classical" load-bearing masonry walls

Although many new techniques and systems of elements have been developed, the classic approach—simple bricklaying of solid masonry walls supporting cast-in-situ reinforced concrete slabs—should not be overlooked. In 1965 Klas Anshelm designed the office building shown in Figs. M.28 and M.29, one of a number of similar projects [M.30]. The walls were 37 cm (15 in.) thick, and the concrete slabs about 20 cm (8 in.) thick (Figs. M.30 and M.31). The clean structural and architectural system provides strength and acoustical and thermal insulation in the same structural elements, while still allowing the possibility of change in use in the future. The building system was claimed to be less expensive than several other systems considered for this project.

Figure M.29 ☐ Photograph of office building for Sydkraft [M.30].

M.10 ☐ Concluding remarks

A fascinating multitude of blocks, bricks, and tiles has been available for a long time. Masonry units, hollow or solid, have been made in many different materials, sizes, colors, and patterns, permitting creation of an almost unlimited number of different brick work appearances. However, the load-bearing capacity of masonry has not been fully utilized in the past. In few buildings has the masonry been stressed to any appreciable extent.

Certainly structural masonry has its dangers, so that it must be carefully used. Due to poor planning, poor workmanship and control, and the wide local differences in material properties, such phenomena as shrinkage, creep, and anisotropy have been hard to predict and to master. The lack of research in the field is also a hindrance. As an example, a good failure criterion for masonry has long been wanted, but is not yet fully established. We can, however, already foresee consistently high compressive

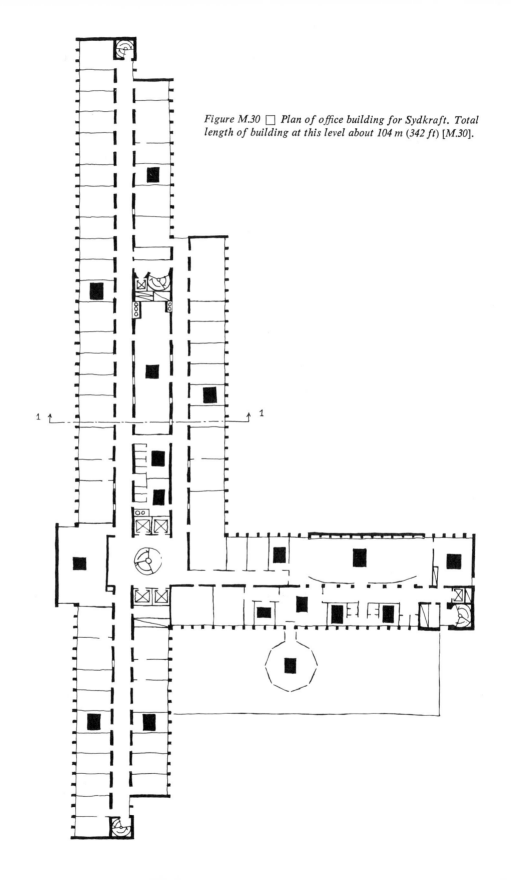

Figure M.30 □ *Plan of office building for Sydkraft. Total length of building at this level about 104 m (342 ft) [M.30].*

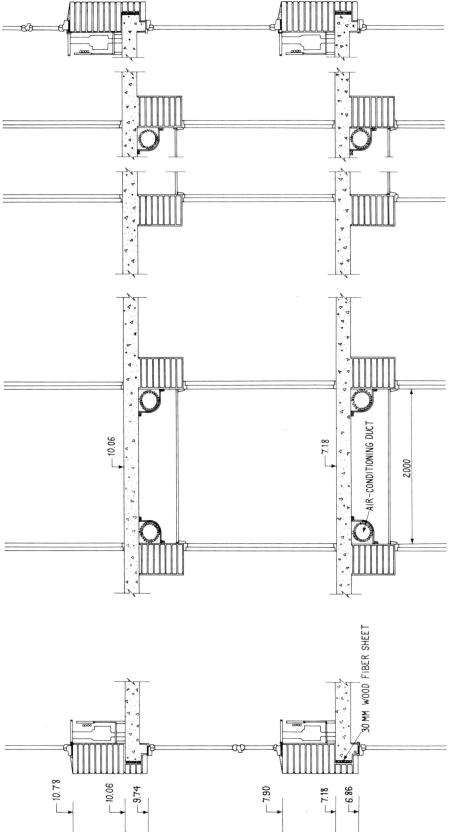

Figure M.31 □ Typical section of office building for Sydkraft (only one story is shown) [M.30].

strength of specially designed masonry units which, in combination with small tolerances and good control of the brick laying process, will result in very high compressive masonry strengths.

When the need for engineering, planning, and control is recognized at all levels of the building process, a fascinating profusion of masonry structures can be anticipated.

References for Chapter M

M.1 ☐ DIN 1053, November, 1962 (German standard).

M.2 ☐ British Standard Code of Practice CP 111, 1964.

M.3 ☐ Svensk Byggnorm 67, BABS, 1967 (Swedish building code).

M.4 ☐ SIA 113, 1965 (Swiss technical standards).

M.5 ☐ Structural Clay Products Institute: "Recommended Building Code Requirements for Engineered Brick Masonry." Washington, D.C., May, 1966.

M.6 ☐ National Concrete Masonry Association: "Recommended Building Code Requirements for Engineered Concrete Masonry." Arlington, Virginia, 1967.

M.7 ☐ American Standard Building Code Requirements for Masonry, U.S. Department of Commerce, NBS Misc. Publ. 211, 1954.

M.8 ☐ Building Code Requirements for Reinforced Masonry, NBS Handbook 74, Superintendent of Documents, 1960.

M.9 ☐ Gross, J,. and Dikkers, R.: "Building Code Requirements Relating to Loadbearing Brick Masonry." In *Designing, Engineering, and Constructing with Masonry Products*, edited by Dr. Franklin Johnson, copyright 1969, Gulf Publishing Company, Houston, Texas. Used by permission.

M.10 ☐ Bradshaw, R., and Foster, B.: "An Assessment of the British Design Methods of Calculated Brickwork." In *Designing, Engineering, and Constructing with Masonry Products*, edited by Dr. Franklin Johnson, copyright 1969, Gulf Publishing Company, Houston, Texas. Used by permission.

M.11 ☐ ASTM Specifications for Concrete Masonry Units: C55, C90, C129, C139, C140, C145, C426, C427.

M.12 ☐ ASTM Specifications for Brick and Tile: C34, C56, C62, C126, C212, C216.

M.13 ☐ ASTM Specifications for Mortar for Unit Masonry: C5, C91, C144, C150, C175, C205, C207, C270, C358.

M.14 ☐ Portland Cement Association: "Concrete Masonry Handbook." PCA, Skokie, Illinois, 1951,

M.15 ☐ Caravaty, R. D., and Plummer, H. C.: "Principles of Clay Masonry Construction: Student's Manual." Structural Clay Products Institute, Washington, D.C., 1960.

M.16 ☐ National Concrete Masonry Association: "Concrete Masonry Foundation Walls." NCMA, CM 131, Arlington, Virginia, 1961.

M.17 ☐ Curtin, W. G., and Hendry, A. W.: "The Design and Construction of Slender Wall Brickwork Buildings." In *Designing, Engineering, and Constructing with Masonry Products*, edited by Dr. Franklin Johnson, copyright 1969, Gulf Publishing Company, Houston, Texas. Used by permission.

M.18 ☐ Rowley, Algren, and Carlson: "Thermal Properties of Concrete Construction, Part 1." In Heating, Piping, and Air Conditioning, January, 1936.

M.19 ☐ Rowley, Algren, and Lander: "Thermal Properties of Concrete Construction, Part 2." In Heating, Piping, and Air Conditioning, November, 1936.

M.20 ☐ Haller, J. Paul: "Hochhäuser in Basel." Die Ziegelindustrie, Nr. 15, 1953.

M.21 ☐ Proceedings of the First National Brick and Tile Bearing Wall Conference. SCPI, Washington, D.C., 1965.

M.22 ☐ Stabilini, L.: "Research and Tests Pertaining to the Use of Baked Clay Products in Walls and Ribbed-Slab Fillers." Topic II, General Reports, Rilem Symposium, Milan, 1962. RILEM Bulletin, New Series, No. 20, September, 1963.

M.23 ☐ Haller, J. Paul: "Mauerwerk in Ingenieurbau." Schweizerische Bauzeitung, 83 Jahrgang, Heft 7, S. 103, February 18, 1965.

M.24 ☐ SCPI: "Contemporary Brick Bearing Walls." Case Study, February–March, 1967.

M.25 ☐ Holmes, T. L.: "Masonry Building in High Intensity Seismic Zones." In *Designing, Engineering, and Constructing*

with Masonry Products, edited by Dr. Franklin Johnson, copyright 1969, Gulf Publishing Company, Houston, Texas. Used by permission.

M.26 □ Nevander, L. E.: "Tekniska Egenskaper hos Isolerade Hålmurar av Tegel." Meddelande Nr. 23, Institutionen för Byggnadsteknik KTH, Stockholm, 1961.

M.27 □ Lättbetong, A. B.: "Lättbetonghandboken." Stockholm, 1965.

M.28 □ Courtesy of A. B. Lättbetong.

M.29 □ Elgenstierna, R.: "Tegelelement." Tegel, No. 2, Stockholm, 1967.

M.30 □ Courtesy Klas Anshelm, Arch. SAR, Lund, Sweden.

Author index

Subject index